A REVIEW OF THE NORTH AND CENTRAL AMERICAN SPECIES OF *PARAVILLA* PAINTER

(Diptera: Bombyliidae)

A Review of the North and Central American Species of *Paravilla* Painter

(Diptera: Bombyliidae)

by Jack C. Hall

UNIVERSITY OF CALIFORNIA PRESS
Berkeley • Los Angeles • London

UNIVERSITY OF CALIFORNIA PUBLICATIONS IN ENTOMOLOGY

Volume 92

Issue Date: January 1981

UNIVERSITY OF CALIFORNIA PRESS
BERKELEY AND LOS ANGELES, CALIFORNIA

UNIVERSITY OF CALIFORNIA PRESS, LTD.
LONDON, ENGLAND

ISBN: 0-520-09625-8

LIBRARY OF CONGRESS CATALOG CARD NUMBER: 80-19994

PRINTED IN THE UNITED STATES OF AMERICA

Library of Congress Cataloging in Publication Data

Hall, Jack Clayton.
A review of the North and Central American species of Paravilla Painter (Diptera--Bombyliidae)

(University of California publications in entomology; v. 92)
Bibliography: p.
Includes index.
1. Paravilla--Classification. 2. Insects--Classification. 3. Insects--North America--Classification. 4. Insects--Central America--Classification. I. Title. II. Series: California. University. University of California publications in entomology; v. 92.
QL537.B65H35 595.77'1 80-19994
ISBN 0-520-09625-8

Contents

Acknowledgments

A special debt of gratitude is owed to the late R. H. Painter and his wife Elizabeth for their painstaking work over the years in collecting, mounting, and preserving numerous Bombyliidae. The Painter collection served as the basis for the present study. Almost all of the new species described herein, particularly those species occurring in Mexico, are the results of the Painters' efforts.

Others from whom specimens were borrowed are: D. W. Webb, Illinois Natural History Survey; M. Wasbauer, California Department of Agriculture; L. Knutson, U.S. Department of Agriculture, Agricultural Research Service, U.S. National Museum; P. H. Arnaud, Jr., California Academy of Sciences; W. J. Hanson, Utah State University; F. F. Hasbrouck, Arizona State University; E. I. Schlinger, University of California at Berkeley; F. G. Werner, University of Arizona; G. Radcliff, University of Nebraska; G. Byers, University of Kansas; L. Pechuman, Cornell University; W. J. Turner, Jr., Washington State University; H. V. Weems, Florida State Department of Agriculture; H. J. Teskey, Canadian National Collection; P. Wygodzinsky, American Museum of Natural History; J. Scott, Museum of Comparative Zoology, Harvard University; R. Schuster, University of California at Davis; E. Ordway, University of Minnesota.

A special word of thanks to K. V. G. Smith, British Museum (Natural History), and M. Beier, Vienna Naturhistoriche Museum, for the loan of holotypes.

INTRODUCTION

The genus *Paravilla* Painter belongs to the tribe Villini, sub-family Exoprosopinae. The flies are known only from the Western Hemisphere and are known to be parasitoids of solitary bees. This study treats the fifty-three species that occur north of Colombia. Of these 27 are described as new. Most species are found in the southwestern United States and northern Mexico; only three species are known from east of the Mississippi River. In South America the genus is represented by two described species (Painter et al. 1978).

Paravilla - along with *Chrysanthrax*, *Thyridanthrax*, *Hemipenthes*, *Diochanthrax*, *Rhynchanthrax* and others - was considered by previous authors as a subgenus of *Villa* Lioy. The historical background of the various components of *Villa* have been discussed and documented by a number of authors (Bezzi 1924; Engel 1932-37; Painter 1933; Hesse 1956; Bowden 1964; Hull 1973; Hall 1970, 1976), and need not be repeated here. Suffice it to say that the European and American authors have differed on the placement of the components of *Villa*, much to the consternation of beginning students.

In 1970, in a study of *Thyridanthrax* I gave a list of possible broad methods for dealing with *Villa,* and decided to follow the subgeneric concepts of previous American authors. Additional study, coupled with Hull's (1973) work on the world genera of Bombyliidae, has created a different attitude. It is now my opinion that the eight subgenera of the old genus *Villa,* i.e., *Villa, Thyridanthrax, Chrysanthrax, Hemipenthes, Paravilla, Diochanthrax, Rhynchanthrax,* and *Paranthrax,* are valid genera. *Exhyalanthrax* is considered a subgenus of *Thyridanthrax* (Bowden 1975). *Hyalanthrax* (listed by Hall 1976, p. 253), is synonymous with *Villa.* The various groups are distinct, though a few problematic species do exist that share characteristics of two or more genera. Future studies will require either erection of new genera to include these species or enlargement of the limits of the present genera.

The basis for the present study is found in Coquillett's 1887 and 1894 papers, Osten Sacken's 1886 study, and Painter and Painter's 1962 study on the types of North American Bombyliidae in European museums. For the most part the species of *Paravilla* have received no attention, except in catalogs, since they were originally described.

While attempting to separate the various species of *Paravilla,* it became apparent that the conventional sorting characters - primarily, color - would not always suffice. In a number of species, especially those with wide distribution, e.g., *P. syrtis* (Coquillett), the variability in color and color patterns was so great as virtually to preclude diagnosis. As a consequence, in making preliminary separations I have placed a good deal of emphasis on the configuration of the male genitalia. Some species, e.g., *P. edititoides* (Painter) and *P. fulvicoma* (Coquillett), cannot be distinguished with confidence by external characters but can easily be separated on the basis of the genitalia. (In some cases, of course, unrelated species exhibit the same, or almost the same, configuration of the genitalia while their external characteristics are quite different; determinations cannot be made, therefore, on the basis of genitalia alone.)

In the descriptions, the dimensions given for the first two antennal segments are taken from a dorsal view. The third antennal segment is described from a lateral view. Unless stated otherwise, the genitalia are described from a lateral view.

Because of the relatively unknown status of the genus *Paravilla* in South America the two described species, *P. stheno* (Wiedemann) from Brazil and *P. villica* (Philippi) from Chile, have been omitted from this study. *P. stheno* is unknown to me and the type specimen, located in the Copenhagen Museum, is unavailable. Painter and Painter (1974, p. 280) redescribed the type of *P. stheno* but from that redescription it is unclear whether *P. stheno* belongs to *Paravilla*. It is closely related to the genus *Neodiplocampta* Curran and when the type becomes available for study it may be found to belong to that genus. *P. villica* has been described recently

(Hall 1976) and should present no problem in identification.

The bee-fly fauna of the Neotropical region is poorly known, except for Chile, and any attempt to include the described and undescribed species (I have seen several of the latter) would entail research well beyond the scope of the present work.

GENUS *PARAVILLA* (PAINTER)

Villa (Paravilla) Painter, 1933: 10. Genotype *edititoides* Painter, (orig. des.). *Paravilla* (Painter), Curran, 1934: 195; Hull, 1973: 391.

This genus is characterized and distinguished from the other members of the tribe Villini by the following characteristics:

Face projecting conically, in some species quite strongly so; third antennal segment either long or short, conical at base, apical portion styliform; arista, when present, apical; proboscis short, at most, narrow labellae projecting beyond oral margin; fore tibia with bristles; fore tarsus not modified, fore claws almost as large as claws on other legs; wing infuscated or hyaline; vestiture of hair and narrow scales; male genitalia with lower apical corner of epandrium (in lateral view) rounded, truncate, pointed or prolonged posteriorly; epiphallus (in lateral view) straight or humped, apex curved up or recurved, (in dorsal view) broad, concave or convex, tonguelike to spatulate; aedeagus (in lateral view) short and broad or long and narrow, with gonopore almost always ventral. Parasites of ground-nesting bees.

The following key to species includes many of the most common variants. As a consequence, many species are keyed in several places.

Key to the Species of *Paravilla* of North and Central America

1. Venter of thorax and abdomen entirely black pilose.. 2
   Venter of abdomen and/or pleural area of thorax not entirely black pilose 8
2. Post-alar bristles pale 3
   Post-alar bristles black 7
3. Fifth abdominal tergite not entirely black tomentose, sides at base with yellowish scales 4
   Fifth abdominal tergite entirely black tomentose 5
4. Dorsum of abdomen with black tomentose crossbands on segments two, three, and four, segments six and seven silver-white tomentose. Mexico .... *eminens*, n. sp.
   Dorsum of abdomen with at most median triangles of black tomentum on segments two, three, and four at most seventh segment only with white tomentum. Mexico north to Colorado .... *extremitis* (Coq.)
5. Scutellar bristles predominantly or entirely black... 6
   Scutellar bristles yellowish, one to two median

bristles sometimes black .... *extremitis* (Coq.)

6. Scutellum with a small, median basal spot of black tomentum; wing usually with basal half or more of axillary cell infuscated; color not reaching posterior margin of axillary cell. Mexico .... *nigriventris*, n. sp.

Scutellum entirely yellowish tomentose: wing usually with color only in axillary cell at extreme base and along vein 2A. Central America north to Arizona and New Mexico .... *consul* (O. S.)

7. Abdomen with two basal segments bright-orange or fulvous pilose, rest of abdomen black pilose and tomentose. Western United States and Mexico .... *spaldingi* (Paint.)

Abdomen with two basal segments white to pale-yellowish pilose .... *consul* (O. S.)

8. Wing hyaline or nearly so, at most with light-brown color in costal and basal portion of first basal cells and on posterior and r-m crossveins, no color in discal cell 9

Wing well marked with brownish color, both basal cells usually colored, as well as anal cell and basal portion at least of discal cell, or wing hyaline with dark areas surrounding r-m and posterior crossveins, and discal cell colored at base 16

9. Wing with three submarginal cells 10
   Wing with two submarginal cells 11
10. Side of second abdominal segment with black hair at least on posterior apical corner. Southern California .... *tricellula* (Cole)
   Side of second abdominal segment pale-yellow pilose, without black hair on posterior corner. Southwestern United States and northern Mexico .... *winburni*, n. sp.
11. Femora (except sometimes apices) black 12
   Femora fulvous or dusky, not black 13
12. Abdominal dorsum with black tomentum in middle of segments two to six; hair and tomentum on body dark fulvous or dark yellow. Midwestern United States .... *nigronasica* (Paint.)
   Abdominal dorsum without black tomentum; hair and tomentum on body pale yellow to whitish. Southern California to New Mexico .... *nigrofemorata* Hall
13. Abdominal dorsum without black tomentum 14
   Abdominal dorsum with black tomentum on second segment, sometimes on third and fourth segments also. South Dakota to Texas .... *xanthina* (Paint.)
14. Dorsal abdominal tomentum pale yellow; white scales, if present, on last two apical segments; mid and hind femora

without black scales 15

Dorsal abdominal tomentum white or at least predominantly whitish; black scales on mid and hind femora. Deserts of California, Arizona, and New Mexico .... *mercedis* (Coq.)

15. Wing completely hyaline, without spots of color on crossveins. Southwestern United States, northern Mexico, Baja California .... *flavipilosa* (Cole)

Wing hyaline, but with spots of color on r-m and posterior crossveins and at base of third posterior cell. California, Arizona, northern Mexico .... *albicera*, n. sp.

16. Wing with three submarginal cells 17

Wing with two submarginal cells 18

17. Side of second abdominal segment with black hair at least on posterior corner; legs fulvous .... *tricellula* (Cole)

Side of second abdominal segment without black hair; legs dusky or at least not bright fulvous .... *winburni*, n. sp.

18. Wing with a spot of color at base of vein R_4 and usually on crossvein at base of second posterior cell, color often quite faint or discal cell almost entirely colored (best seen in indirect light) 19

Wing without spot of color at base of vein R_4, although there may be faint color on crossvein at base of second posterior cell; color in discal cell usually not extending beyond r-m crossvein 37

19. Legs entirely fulvous or, at most, femors slightly dusky 20

Legs mostly or entirely black (especially femora) or legs so densely covered with black scales and setulae as to appear black 28

20. Abdominal dorsum with black tomentum either as crossbands or as small median triangles, sometimes on fourth segment only 21

Abdominal dorsum without black tomentum 25

21. Margin of color in wing delineated sharply 22

Margin of color in wing evanescent, not sharp. Central California coast .... *fumosa*, n. sp.

22. Basal half or more of wing infuscated, color not forming bands; color in first posterior cell not reaching wing margin 23

Color in wing forming three distinct bands, one basal, one median, one apical; color in first posterior cell reaching margin of wing. Mexico .... *castanea* (Jaenn.)

23. Propleuron with white or pale-yellow hair, black hair absent 24

Propleuron with at least a few black hairs present among the pale yellow. Mexico and Central America .... *eurhinata* (Big.)

24. Apical half of abdomen fulvous in ground-color; color in discal cell reaching r-m crossvein. Southern California, Baja California .... *splendida*, n. sp.

Abdominal ground-color black, color in discal cell not reaching r-m crossvein except as a border along vein M_1. Western United States, Canada to Arizona .... *vigilans* (Coq.)

25. Margin of color in wing delineated sharply 26

Margin of color in wing evanescent, not sharp 27

26. Side of abdomen with black hair on second or third and following segments .... *vigilans* (Coq.)

Side of abdomen without black hair; black hairs, if present, on apical segments and few in number. New Mexico, Arizona, Colorado .... *emulata* (Paint.)

27. Side of occiput with white scales;
small species, 6-10 mm. in length.
California .... *fumida* (Coq.)
Side of occiput with yellow scales;
large species, 12-15 mm. in length.
Western United States, Canada to
Mexico .... *montivaga*, n. sp.

28. Side of abdomen with black hair on
posterior apical corner of second
segment 29
Side of abdomen without black hair
anterior to the third segment 33

29. Scutellar bristles black 30
Scutellar bristles fulvous or pale 31

30. Venter of abdomen predominantly white
tomentose; dark scales, if present,
on the apical segments; lower apical
corner of epandrium pointed. Southern
California and Arizona .... *imitans*, n. sp.
Venter of abdomen mixed black and
white tomentose; lower apical corner of
epandrium rounded. California, Arizona,
Colorado, Wyoming .... *macneilli*, n. sp.

31. Apical fourth of anal cell hyaline or
at least color not reaching hind
margin of wing 32
Anal cell colored completely, color
reaching hind margin of wing; color

often present only along veins Cu_2 and 2A .... *macneilli*, n. sp.

32. Wing with dark color surrounding crossvein at base of second posterior cell. Arizona south throughout Mexico .... *epheba* (O. S.)

Wing without a spot of color on crossvein at base of second posterior cell; color, if present, very faint. Western United States, Canada to Mexico .... *borea*, n. sp.

33. Scutellar bristles pale yellow to fulvous, at most one or two black bristles present 34

Scutellar bristles black, at least in middle along posterior margin .... *macneilli*, n. sp.

34. Abdominal segments six and seven with yellowish or white scales 35

Abdominal segments six and seven with silver-white scales 36

35. Color in wing not extending beyond r-m crossvein in discal cell except as a border to vein M_1; margin of color delineated sharply .... *borea*, n. sp.

Color in wing reaching base of second posterior cell; discal cell almost entirely colored; margin of color evanescent. Southwestern United States .... *albata,* n. sp.

36. Side of abdomen with black hair on third and following segments ...... *leucothoa* (Wied.)

Side of abdomen without black hair: black hairs, if present, few in number and commencing on the fifth segment. Mexico .... *mexicana,* n. sp.

37. Second abdominal segment with black hair on posterior lateral corner (one or two hairs present are significant) 38

Second abdominal segment without black hair on side 57

38. Second abdominal segment with black tomentum in the form either of a crossband or a small medial spot 39

Second abdominal segment without black tomentum (a few dark brown scales may be present but are to be disregarded) 50

39. Scutellar bristles black at least in middle of posterior margin, or bristles at least not all light yellowish 40

Scutellar bristles yellow or fulvous, not dark 44

40. Tomentum on scutellum black, pale yellow, or fulvous, never pure

white 41

Tomentum on scutellum pure white. Southwestern United States and Mexico .... *cunicula* (O. S.)

41. Face and first antennal segment black; face sometimes fulvous along oral margin and below 42

Face and first antennal segment fulvous; face sometimes black in middle above. East of Rocky Mtns., from Canada to Texas and Florida .... *separata* (Wlk.)

42. Fourth abdominal segment with pale scales, white to yellow 43

Fourth abdominal segment entirely black tomentose. Florida .... *floridensis*, n. sp.

43. Fifth abdominal segment entirely black tomentose. Southwestern United States and northern Mexico .... *diagonalis* (Lw.)

Fifth abdominal segment yellowish tomentose, at most a few black scales sometimes present in middle on posterior margin. Texas and Mexico .... *opaca*, n. sp.

44. Second basal cell entirely infuscated or basal half subhyaline; if nearly hyaline, then color on posterior

crossvein connecting with color at base of discal cell 45

Second basal cell hyaline to subhyaline, at least lighter in color than isolated spot on posterior crossvein. Southwestern United States and Mexico .... *acutula*, n. sp.

45. Black tomentose crossband on third abdominal segment complete or incomplete; if a small spot, at least expanded laterally 46

Black tomentum on third abdominal segment, if present, in the form of a small spot not expanded laterally; black tomentum often absent 49

46. Side of mesonotum between humeral callus and wing base entirely pale-yellow to white pilose 47

Side of mesonotum between humeral callus and wing base, as well as upper part of mesopleuron, with scattered black hair and/or bristles mixed with the pale hair. Western United States and Mexico .... *parvula*, n. sp.

47. Legs black or dusky; abdominal dorsum normally with black and yellow tomentum only 48

Legs fulvous, at most fore femora dusky toward base; abdomen usually with white, fulvous, and black scales. West of Rocky Mtns. .... *syrtis* (Coq.)

48. Discal cell infuscated to r-m crossvein; lower apical corner of epandrium prolonged .... *opaca*, n. sp.

Discal cell not infuscated to r-m crossvein except as a border to vein M_1; lower apical corner of epandrium pointed. Western United States, Canada to Mexico .... *hulli*, n. sp.

49. Lateral margin of abdomen with scattered black hair, not in the form of definite tufts. Southwestern United States and Mexico .... *deserta*, n. sp.

Lateral margin of abdomen with definite tufts of black hair at least on posterior corners of segments two, three, and four .... *parvula*, n. sp.

50. Side of mesonotum with black bristle-like hair between humeral callus and wing base 51

Side of mesonotum without black bristle-like hair between humeral callus and wing base 52

51. Axillary cell mostly hyaline, color

in middle along vein 2A only .... *parvula*, n. sp.

Axillary cell nearly entirely colored, at most apex hyaline. East of Rocky Mtns., Canada to Louisiana .... *palliata* (Lw.)

52. Second basal cell infuscated, sometimes with a subhyaline spot towards the apex 53

Second basal cell hyaline or nearly so, colored along bordering veins only. Western United States and Mexico .... *acutula*, n. sp.

53. Tomentum on scutellum and second abdominal segment predominantly pale yellow to yellow 54

Tomentum on scutellum and second abdominal segment predominantly white. Western United States and Mexico .... *deserta*, n. sp.

54. Third antennal segment one and one-half times length of segments one and two combined; body pile pale yellow to white 55

Third antennal segment twice length of segments one and two combined; body pile normally light fulvous. Western United States and northern Mexico .... *fulvicoma* (Coq.)

55. Black hair on mesonotum relatively dense, conspicuous; aedeagus short, curved 56
    Black hair on mesonotum short, sparce, and inconspicuous; aedeagus straight, more than half length of epiphallus. California and Arizona .... *parasitica*, n. sp.

56. Fulvous scales on abdomen at least across posterior margins of segments two and three; face usually at least partially black .... *hulli*, n. sp.
    Tomentum on abdomen pale yellowish, without fulvous scales; face entirely fulvous. Nebraska to Mexico, west to Arizona .... *pallida*, n. sp.

57. Second abdominal segment with black tomentum, in the form of a median spot or a crossband 58
    Second abdominal segment without black tomentum, although dark-brown scales sometimes present 70

58. Fourth abdominal segment with black tomentum, only a few black scales sometimes present medially 59
    Fourth abdominal segment without black scales 62

59. Black tomentum on fourth abdominal segment in the form of a cross-

band on posterior margin of segment 60

Black tomentum on fourth abdominal segment at most forming small medial spot 61

60\. Basal half of first abdominal segment white pilose and tomentose; color in wing dark brown .... *cunicula* (O. S.)

Basal half of first segment pale-yellow pilose and tomentose; color in wing pale brown. Southwestern states, Texas to Nevada .... *texana*, n. sp.

61\. Face and first antennal segment black; pile and tomentum at base of abdomen pale yellowish. Texas and Mexico ..... *opaca*, n. sp.

Face and first antennal segment fulvous; pile and tomentum at base of abdomen pure white .... *syrtis* (Coq.)

62\. Legs with at least fore femora black or dusky, sometimes faintly so 63

Legs, except tips of tarsi, entirely fulvous 67

63\. First antennal segment and face black 64

First antennal segment and face fulvous 65

64\. Axillary cell completely hyaline; base of third antennal segment long conical, equal in length to apical styliform portion .... *winburni*, n. sp.

Axillary cell at least partially colored; base of third antennal segment short conical, apical styliform portion much longer than conical base .... *opaca*, n. sp.

65. Second abdominal segment without fulvous scales; tomentum white, black or brown scales often present 66

Second abdominal segment with fulvous scales between white and black scales .... *syrtis* (Coq.)

66. Axillary cell hyaline. Southern California .... *perplexa* (Coq.)

Axillary cell infuscated on basal third or more .... *deserta*, n. sp.

67. Ground-color of venter of abdomen bright fulvous, at most first two segments darkened 68

Ground-color of venter of abdomen black, with posterior margins of segments narrowly fulvous or at most with three apical segments fulvous 69

68. Wing with anal cell colored to apex; second basal cell hyaline

to subhyaline .... *lacunaris* (Coq.)

Color in anal cell not reaching posterior margin of wing; second basal cell same color as first basal cell .... *syrtis* (Coq.)

69. Scutellum white tomentose at least across basal half .... *deserta*, n. sp.

Scutellum yellow tomentose; at most, sides at base white tomentose .... *perplexa* (Coq.)

70. Legs black, at least femora black or dusky, sometimes at base only 71

Legs fulvous, at most basal third of fore femur occasionally dusky 78

71. Front with black hair or mixed black and fulvous hair, never entirely fulvous 72

Front with fulvous hair. California, Nevada, Arizona .... *albicera*, n. sp.

72. Wing more or less uniformly colored on basal half, without prominent darker areas on crossveins 73

Wing lightly colored, with prominent darker areas on crossveins. California, Arizona, Mexico .... *cinerea* (Cole)

73. Abdominal dorsum with pure-white scales as well as fulvous and/or yellow scales 74

Abdominal dorsum without pure-white

scales. Mexico .... *painterorum*, n. sp.

74. Antenna black 75
    Antenna with at least first segment
    fulvous 76

75. Color in wing reaching to base of
    second posterior cell; outline
    of color not delineated sharply.
    Southwestern United States deserts ....
    .... *albata*, n. sp.
    Color in wing not reaching beyond
    r-m crossvein except possibly as
    a border to vein M_1; color mar-
    gined sharply .... *syrtis* (Coq.)

76. Scutellar bristles fulvous or
    yellow 77
    Scutellar bristles black .... *cunicula* (O. S.)

77. Fourth abdominal segment black
    tomentose. Deserts of California
    and Arizona .... *aridula*, n. sp.
    Fourth abdominal segment fulvous
    to yellow tomentose .... *deserta*, n. sp.

78. Abdominal dorsum white to pale-
    yellow pilose at base, some-
    times a few short black hairs
    present on posterior margins
    of apical segments 79
    Abdominal dorsum either entirely
    yellow pilose at base or, if

almost entirely white pilose, then with abundant black hairs present other than those across posterior margins of segment two to apex 82

79. Legs entirely fulvous, tips of tarsi black 80

Legs fulvous, with basal half of femora and tips of tarsi dusky .... *fulvicoma* (Coq.)

80. Side of abdomen with black hair on apical corners of at least the third and following segments; wing without a spot of color at base of second posterior cell .... *deserta*, n. sp.

Side of abdomen white, without black hair; wing with or without a faint spot of color at base of second posterior cell 81

81. Wing dark brown on basal half, second cell entirely infuscated. Western United States, Canada to Arizona .... *vigilans* (Coq.)

Wing with light-tan infuscations, second basal cell subhyaline to hyaline .... *albicera*, n. sp.

82. Wing with second basal cell hyaline or nearly so; if colored at all,

lighter than color in first basal cell 83

Second basal cell colored, normally the same color as in first basal cell 84

83. Front black pilose: legs usually with black scales on apical halves of anterior surfaces of femora. Baja California .... *lucida*, n. sp.

Front with black and fulvous hair; legs without black scales on femora. California, Arizona, Nevada, Mexico .... *californica*, n. sp.

84. Side of mesonotum between humeral callus and wing base with white or yellowish hair 85

Side of mesonotum between humeral callus and wing base with scattered black hair mixed with the pale-yellowish hair .... *palliata* (Lw.)

85. Propleuron and coxae black; styliform portion of third antennal segment longer than conical basal portion 86

Propleuron and coxae at least partially fulvous; styliform portion of third antennal segment usually shorter than conical basal portion. Central California and Idaho .... *apicola* (Cole)

86\. Front with black hair or with mixed black and fulvous hair 87

Front entirely fulvous pilose or, at most, with a few scattered black hairs present; body pile and tomentum pale yellowish. California, Arizona, Mexico .... *inatra*, n. sp.

87\. Aedeagus straight, long and narrow, more than half length of epiphallus: species most common in spring and early summer (May to July) .... *fulvicoma* (Coq.)

Aedeagus short, broad and curved downward, less than half length of epiphallus; species most common in later summer and fall (September and October). Western United States, Canada to Mexico .... *edititoides* (Painter)

Paravilla acutula Hall, new species

(Fig. 1)

This species resembles *P. syrtis* but differs in having the second basal cell in the wing hyaline to subhyaline, yellowish pile and tomentum, and fulvous legs, except for the coxae and tips of the tarsi. The lower apical corner of the epandrium is prolonged and acuminate in the *P. acutula* male.

Male. - Black; first two antennal segments, face except middle above, legs except tips of tarsi fulvous.

Front and face with black hair and pale-yellow to whitish scales, latter dense between and above antennae. First antennal segment more than twice length of second segment, expanded slightly on inner apical margin; second segment as long as wide; both basal segments with black hair; third segment about two and one-half times longer than first segment, base short conical, apical styliform portion one and one-half times as long as base. Proboscis with tip of labellae extending beyond oral margin. Palpus with dark-fulvous to black hair. Occiput whitish tomentose, hair at vertex whitish; cervical fringe whitish to pale yellowish.

Mesonotum pale-yellow pilose and tomentose, black hair on disk, white hair along lateral margin, longer scales on posterior half of mesonotum whitish, bristles yellow. Pleura pale-yellow pilose, hair on posterior half of mesopleuron white, more or less contiguous with that above root of wing. Coxae with yellow and black hair and flavotestaceous scales. Femora with black hair and yellow scales, black scales on anterior surfaces of all femora, tibiae with yellow scales, black setulae and bristles, tarsi with black setulae.

Halter stem flavotestaceous, knob white. Squama flavotestaceous, with fringe of yellow scales. Scutellum with black hair, yellow hair present at sides near base, tomentum whitish; bristles yellow.

Basicosta with yellow scales and black setulae. Wing hyaline, brownish color in costal, basal half of marginal, first basal cells, and extreme base of wing, with more or less isolated spots at base of discal cell, bases of third and fourth posterior cells; second basal cell hyaline to subhyaline except at extreme base; vein at base of second posterior cell slightly sinuous.

Abdominal dorsum pale-yellow pilose, scattered black hair across segments two to six, tomentum yellowish, lighter across bases of segments two, three, and four, narrow black tomentose bands across middle of segments two and three, small black tomentose spot in middle of fourth segment, black tomentum not dense; lateral margin of abdomen pale-yellow to whitish pilose, tufts of black hair on posterior corners of segments two and three, scattered black hair mixed with pale hair on rest of lateral margin from segments four to seven. Venter of abdomen with whitish hair on segments one, two, and sides of three, rest with scattered black hair, tomentum flavotestaceous.

Genitalia with epandrium elongate, more than twice as long as median width, lower apical corner prolonged, apex of prolongation acuminate, lower margin of epandrium straight; epiphallus broad, curved slightly, tip curved upward, apex truncate, wrinkled, lower margin convex; in dorsal view, epiphallus broad, apex rounded, sides concave near middle; aedeagus short, less than half length of epiphallus, curved downward; basistylus broad, base rounded, apex pointed bluntly; dististylus short, not hooked strongly at apex.

Female. - Upper half of front with fulvous hair; otherwise essentially as described for male.

Holotype male and allotype. - From State Park, Scott County, Kansas, IX.1.55 (R. H. Painter). Both in U.S. National Museum.

Paratypes. - UNITED STATES: Arizona: Coconino County, 2, The Gap, VIII.28.57 (R. H. and E. M. Painter). Colorado: Logan County, 1, 16 mi. W Sterling, IX.2.64 (E. M. Painter). Kansas: Graham County, 2, VIII.21-30.55 (W. W. Gibson, R. H. Painter); Reno County, 1, sand dunes, Medora IX.20.31 R. H. Painter); Scott County, 29, topotypic, collected with types (R. H. Painter, W. W. Gibson); Sheridan County, 1, VIII.31.55 (R. H. Painter); Smith County, 1, IX.14.12 (F. X. Williams); Thomas County, 1, VIII.31.55 (R. H. Painter). Nebraska: Cherry County, 1, 9 mi. S Nenzel, VIII.34.54 (L. W. Quate); Sioux County, 1, Montrose,

VIII.11.55 (L. W. Quate). Texas: Brown County, 1, Brownwood, VIII.28.31 (R. H. Painter); Mitchell County, 3, Loraine, IX.14.21 (R. H. Painter); Taylor County, 1, Merkel, IX.15.21 (R. H. Painter). MEXICO: Durango: 1, 7 mi. NE Durango, IX.20.59 (R. H. and E. M. Painter).

Paratypes in the collections of the United States National Museum, the University of Nebraska, and the University of California at Riverside.

Paravilla albata Hall, new species

(Fig. 2)

This species is allied to *P. fumosa*, but the wing coloring is more sharply margined posteriorly, the color reaches the base of the second posterior cell, the abdominal pile is white and the legs and antennae are black.

Male. - Black; middle of side of face with a testaceous spot, oral cavity flavotestaceous.

Front and face black pilose and pale-yellow tomentose, scales on front reaching to ocellar tubercle. Antenna with black hair; first segment almost three times length of second segment, expanded on inner apical margin; second segment one and one-half times wider than long; third segment with base long conical, apical styliform portion equal to conical base. Labellae projecting beyond tip of oral margin. Palpus with black hair. Occiput yellow tomentose, a few black hairs at vertex; cervical fringe pale yellow.

Disk of mesonotum with short black hair, mesonotum otherwise pale-yellow pilose and tomentose, the latter dense, hair a little lighter above root of wing, bristles yellow.

Pleura pale-yellow pilose. Coxae and femora with black hair and scales, mid and hind femora with yellowish scales on dorsal surfaces, hind tibia with black scales, yellowish scales on dorsal surface, other tibiae with black setulae and bristles.

Halter stem flavotestaceous, knob creamy white. Squama brownish, with fringe of white scales. Scutellum with black hair, white hair present at sides near base; bristles yellow.

Basicosta with yellowish scales and black setulae. Basal two-thirds of wing infuscated with brown, outline of color not sharp, extending diagonally basad from tip of subcostal vein to apex of anal cell, both basal cells, anal cells and axillary cells entirely infuscated, discal cell almost entirely colored, vein at base of second posterior cell bordered with brown.

Abdominal dorsum white to whitish pilose and tomentose, black hair across posterior margins of segments two to four, hair at apex of abdomen yellowish; lateral margin of abdomen whitish pilose, a few black hairs present on posterior corners of segments three and four. Venter of abdomen whitish pilose and tomentose.

Genitalia with epandrium not quite twice as long as high, lower apical corner almost square, lower margin straight, epiphallus broad, tip curved upward gently, apex rounded, lower margin bulbous beyond middle; in dorsal view, epiphallus broad, rounded broadly at apex, sides concave medially; aedeagus short, less than half length of epiphallus, curved downward; basistylus boat-shaped, apex bluntly rounded; dististylus long, narrow, hooked apically.

Female. - Except for more abundant and whiter tomentum, female does not differ from male.

Holotype male and allotype. - From Deep Canyon, 3.5 mi. S Palm Desert, Riverside County, California, IV.17.73 (A. B. Tabet). Both in U.S. National Museum.

Paratypes. - UNITED STATES: Arizona: Yuma County, 1, Liquita, IV.9.47 (A. L. Melander). California: Inyo County, 2, Eureka Valley, V.4.77 (J. C. Hall); Riverside County, 16, topotypic, collected with type; 1, 10 mi. E Mecca, IV.13.63 (E. M. Fisher); 1, Thousand Palms Canyon, IV.5.66 (J. C. Hall); San Bernardino County, 3, Granite Pass, IV.27.68 (E. I. Schlinger); San Diego County, 2, Borego, V.4.45, IV.26.55 (A. L. Melander, P. D. Hurd); 2, 5 mi. E Jacumba, III.18.56 (T. R. Haig); 2, Ocotillo, III.3.47, IV.6.47 (G. H. Crickner and J. L. Sperry). Nevada: Nye County, 1, Mercury, Nevada Test Site, IV.28.62. New Mexico: Lincoln County, 3, Picacho, IV.9.10 (F. M. Hull).

Paratypes in the collections of the United States National Museum and the University of California at Riverside.

Paravilla albicera Hall, new species

(Fig. 3)

This species is similar to both *P. winburni* and *P. flavipilosa*. *P. albicera* is separated from *P. flavipilosa* by the light wing infuscations and black hair on the abdomen, and from *P. winburni* by the entire front being pale-yellow pilose.

Male. - Black; lower half of front, first two antennal

segments, face, apices of femora, all of tibiae, posterior margins of apical abdominal segments fulvous.

Front yellow pilose, white tomentose, tomentum not reaching ocellar tubercle; a few black hairs on ocellar tubercle. First two antennal segments with yellow hair, first segment cylindrical, with black hair above, expanded slightly on inner apical margin, twice length of second segment; second segment oval, as wide as long; third segment one and one-half times longer than first two segments combined, tapering rather evenly from broad base to styliform apex. Face with yellow hair and white scales; a few black hairs sometimes present at tip of oral margin. Proboscis not projecting beyond oral margin. Palpus with black hair. Side of occiput with whitish scales, becoming yellow toward vertex; cervical fringe white.

Disk of mesonotum with short, scattered, black hair; anterior and posterior margins with pale-yellowish hair, white hair on lateral margin above root of wing; tomentum dense, yellowish white, bristles yellow. Pleura pale-yellow to white pilose and tomentose. Coxae with yellow hair, tomentum yellowish white; fore and mid femora with inconspicuous yellow hair, absent on hind femur, femora and tibiae with dull-yellow scales, all bristles black.

Halter stem flavotestaceous, knob creamy yellow. Squama flavotestaceous, with fringe of white scales. Scutellum with yellow hair, tomentum pale yellowish, lighter at sides and along posterior margin, dense; bristles yellow.

Basicosta with yellowish scales and black setulae. Wing

hyaline, light yellowish-brown color in costal and subcostal cells, spot at knee of vein R_{2+3} extending into marginal cell; spot on r-m crossvein, bases of third and fourth posterior cells, color diffused along basal half of vein R_5 to r-m crossvein; discal and second basal cells hyaline except for color along enclosing veins; anal cell hyaline.

Abdominal dorsum pale-yellow to whitish pilose; a few scattered, short, black hairs in middle of posterior margins of segments two to four; tomentum pale-yellow to whitish, pure white on segment seven, slightly darker-yellow scales in narrow bands across posterior margins of segments two to five; side of abdomen entirely whitish pilose. Venter of abdomen pale-yellow pilose, white tomentose.

Genitalia with epandrium narrow, more than twice as long as high, lower apical corner prolonged, prolongation broad, about as wide as long, lower margin concave medially; epiphallus straight, narrow, tip curved upward, apex rounded, lower margin straight; in dorsal view, epiphallus broad apically, apex rounded, sides concave near middle; aedeagus half length of epiphallus, curved downward; basistylus narrow, pointed at base, apex bluntly rounded; dististylus long, narrow, hooked apically.

Female. - Except for usual sexual differences, plus more black hair on the abdomen, especially at the sides of segments three to seven, and yellow tomentum on the seventh segment, female does not differ from male.

Variation. - A few specimens have a few black hairs on the front near the ocellar tubercle. The legs may be dark or light, but are always dusky.

Holotype (male) and allotype. - From Ajo, Pima County, VIII.30.58 (R. H. and E. M. Painter). Both in U.S. National Museum.

Paratypes. - UNITED STATES: Arizona: Pima County, 13, topotypic, collected with types; 1, Sells, VIII.31.32 (R. H. Painter). California: Alameda County, 2, hills behind Oakland, V.16.09 (E. C. Van Dyke); Imperial County, 1, Indian Wash, IX.28.71 (J. C. Hall); Lassen County, 1, Halleluja Junction, VII.4.49 (R. H. Howell); Los Angeles County, 1, Tanbark Flat, VI.20.50 (P. D. Hurd); Monterey County, 1, Bradley, IX.16.64 (F. M. Hull); Riverside County, 6, Andreas Canyon, IV.24.54 (J. G. Rozen); 21 Hopkins Well, IV.16.58 (J. W. MacSwain); San Diego County, 2, San Diego, V.15.13 (E. P. Van Duzee); Ventura County, 1, E Anacapa Island, VIII. 22.78 (S. E. Miller). Nevada: Nye County, 2, Mercury, VIII. 26.59, VI.4.61. MEXICO: Chihuahua, 2, 4 mi. S Villa Ahumada, IX.7.62 (R. H. and E. M. Painter).

Paratypes in the collections of the California Academy of Sciences, the University of California at Berkeley and at Riverside, and the United States National Museum.

Paravilla apicola (Cole)

Villa apicola Cole, 1952: 128. (Type in California Academy of Sciences). *Villa (Paravilla) apicola* Cole, Painter and Painter, 1965: 438. *Paravilla apicola* (Cole), Hull, 1973: 392.

The holotype, contrary to the statement of Cole, is a female. This species was reared from the nests of *Diadasia consociata* Timb. (Cole 1952).

P. apicola is most similar to *P. edititoides* and *P. fulvicoma*; it can be separated from either species by the fulvous-colored propleuron and coxae and by the configuration of the third antennal segment - in *P. apicola*, the styliform portion of the third antennal segment is shorter than the rather broad basal portion, while in both *P. edititoides* and *P. fulvicoma* the styliform portion of the third antennal segment is much longer than the swollen basal portion.

Male. - Unknown.

Female. - Small, 7-9 mm. in length. Black; first antennal segment, face, palpi, coxae, and legs except tips of tibiae, abdominal venter, and propleuron fulvous.

Front with mixed black and fulvous hair, the latter, usually predominating; scales dense, dull yellowish. Basal antennal segments with short black hair above, fulvous hair below; first segment twice length of second, not appreciably expanded apically; second segment almost two and one-half times wider than long; third segment wider than second, broad, elongate conical basally, apical styliform part equal to or shorter than enlarged base, basal portion about equal to length of first two segments combined. Face with black hair, some fulvous hairs along oral margin and at sides of face next to eyes; scattered dull-yellowish scales present. Proboscis short; at most, labellae projecting beyond tip of oral margin. Palpus with yellow hair. Occiput with yellow hair and scales; cervical fringe pale yellowish.

Mesonotum pale-yellow pilose and tomentose, side above root of wing a little lighter in color, bristles yellowish.

Pleura pale-yellow pilose and tomentose. Legs, including coxae, yellow pilose and tomentose, femora and tibiae with black bristles.

Halter stem yellowish, knob whitish. Squama flavotestaceous, with fringe of pale-yellow scales. Scutellum with pale-yellow hair and scales; bristles yellowish.

Basicosta with yellow scales and black setulae. Wing colored light brownish-yellow on basal half, hyaline on apical half; color runs diagonally basad from tip of subcostal cell to middle of axillary cell, both basal cells colored uniformly, color in discal cell reaching r-m crossvein and beyond along vein M_1;anal cell colored on basal two-thirds, color reaching posterior margin of wing only at or near base of axillary cell; alula entirely colored, first posterior cell narrowed apically, open in wing margin for a distance less than length of r-m crossvein; medial crossvein at base of second posterior cell sinuous, but not strongly so.

Abdomen dull yellow pilose and tomentose; posterior margins of segments two to six with slightly darker-colored scales. Side and venter of abdomen like the dorsum.

Distribution. - Baja California, Arizona, Texas, California. In California, San Joaquin County, south to San Luis Obispo and Kern counties; the San Joaquin County specimens were reared from nests of *Diadasia consociata* Timberlake. I have also seen seven specimens from 12 mi. W Gooding, Gooding Co., Idaho, VIII.11.77 (A. B. Tabet) that do not differ from Cole's type series.

Paravilla aridula Hall, new species

Fig. 4

A large species (12-15 mm. in length), *P. aridula* is quite unlike most of the congeners but approaches *P. cinerea*. The basal portion of the wing is dark brown; pile and tomentum of the dorsum of the body are predominantly white, black hair present on tergites four and five; the macrochaetae are fulvous.

Male. - Black; basal antennal segments, lower half of front, side of face dark ferruginous to brown.

Eyes separated at narrowest point by slightly more than twice width of ocellar tubercle. Front with short black hair, absent in middle above antennae; tomentum white, not extending to ocellar tubercle. First antennal segment twice length of second, from above a little broadened apically, both basal segments with black hair; third segment longer than first two segments combined, elongate conical, tapering from broad base to apex, at most last fourth may be considered styliform, style small, terminal. Facial (strongly conical) projection reaching beyond tip of antennae, with short black hair and white tomentum, bare next to eye. Proboscis with, at most, labellae projecting beyond oral margin. Palpus with black hair. Occiput white tomentose; cervical fringe white above, pale yellowish below.

Mesonotum with short, black hair mesally; apical hair pale-yellowish, sides whitish, long hair in front of scutellum and above base of wing shining fulvous; tomentum white, dense; bristles fulvous. Pleura pale-yellow to white pilose, white tomentose. Coxae with black hair and long brownish-black scales, femora and tibiae with black scales, fore femur with black hair, absent on other femora, bristles strong black.

Halter stem brown, knob white. Squama fulvous, with fringe of yellowish scales. Scutellum with scattered black hair, a few shining fulvous hairs across base and at side; tomentum white; bristles fulvous.

Broadened base of costa with white scales at base, grading evenly into yellow, then black, apically. Wing with basal half or less infuscated, apical portion hyaline; color extends diagonally basad from apex of subcosta across discal cell to middle of axillary cell, color extending into base of first submarginal cell, basal half of discal cell or less, extreme basal corner of third posterior cell, broadly across base of fourth posterior cell, across basal half of anal cell; first posterior cell narrowed apically, open in wing margin for distance less than length of r-m crossvein; cross-vein closing apex of discal cell at most only slightly sinuous, r-m crossvein slightly beyond middle of discal cell, alula brown with fringe of brownish scales.

Abdominal dorsum white pilose, a few black hairs in middle of posterior margin of segment two, many across posterior half of segment three, all of four and five; tomentum whitish on segments two and three, brownish on four and five, pure white on segments six and seven; lateral margin of abdomen white-haired on segments one to three, six and seven, posterior corner of third segment, all of fourth and fifth segments black-haired. Venter of abdomen white pilose and tomentose; band of black hair on sternites five and six, hair on sternite seven white to pale-yellow.

Genital capsule testaceous, epandrium long, narrow, lower apical corner prolonged, rounded; epiphallus long, straight or nearly so, broad, tip curved upward; aedeagus half length of epiphallus, bent downward, tip acuminate; in dorsal view, aedeagus not broadened apically; dististylus small, sharply hooked apically.

Female - Face, basal antennal segments light fulvous, tomentum on third dorsal abdominal segment yellowish medially, black hair on venter of abdomen on segments four to apex; otherwise as described for male.

Holotype male and allotype. - From Deep Canyon, Riverside County, California IX.7.73 (A. B. Tabet). Both in California Academy of Sciences.

Paratypes. - UNITED STATES: Arizona: Yuma County, 1, 15 mi. E Quartzite, IX.16.59 (P. H. Timberlake); 1, 2.3 mi. W Branda, IX.14.42 (P. H. Timberlake), on *Bebbia juncea.* California: Imperial County, 6, Indian Wash, IX.21.71 (J. C. Hall); Riverside County, 9, topotypic, collected with type; 2, 10 mi. W Blythe, X.2.54 (J. C. Hall); 21, Deep Canyon, IX. 5.63, X.4.63, X.13.63, X.28.63, VIII.30.73, VIII.30-IX.7.73 (E. L. Schlinger, M. E. Irwin, J. D. Pinto, A. B. Tabet), on *Bebbia juncea;* San Bernardino County, 1, Cushenberry Springs, IX.1.36 (P. H. Timberlake), on *Gutierrezia lucida.*

Variation. - Specimens from Indian Wash have the frontal hair yellow rather than black; in the female the brownish-yellow scales on the dorsum of the abdomen may be present on the apical half of the third segment or may be entirely absent. Some specimens may have the legs dark-brownish rather than black. Females appear to vary more in coloration than do the males.

Paravilla borea Hall, new species

Fig. 5

Similar to *P. epheba*. *P. borea* may be separated from *P. epheba* by less extensive wing coloring, faintness of the spot at the base of vein R_4, *P. boreas's* more northern distribution, and male genitalic differences.

Male. - Black; lower half of face, apices of femora, all of tibiae fulvous.

Front and face with black hair and yellow scales. Basal antennal segments with black hair; first segment twice length of second, expanded on inner margin at apex; second segment slightly wider than long; third segment long conical at base, apical styliform portion equal to basal portion; third segment not much longer than two basal segments combined. Proboscis not projecting. Palpus with black hair. Occiput with pale-yellow scales laterally, hair at vertex yellow; cervical fringe yellow.

Mesonotum yellowish pilose and tomentose, tomentum loose, not appressed, scattered, black hair on disk reaching posterior margin, pile whitish above root of wing, bristles yellow. Pleura pale-yellow pilose and tomentose. Coxae with yellow hair and scales, fore and mid coxae with mixed yellow and black hair; femora with yellow scales, fore and mid femora with black hair, hind femur with yellow hair; mid and hind tibiae with yellow scales, all bristles black.

Halter stem testaceous, knob yellowish white. Squama white with fringe of pale-yellow scales. Scutellum pale-yellow pilose and tomentose, on occasion a few black hairs present; bristles yellow.

Basicosta with fulvous scales and black setulae. Wing infuscated on basal half or less; brown color extends from tip of vein R_1 diagonally basad across marginal cell, basal fourth of first submarginal, base of first posterior cell, all of base of discal cell to r-m crossvein, to extreme bases of third and fourth posterior cells, straight across anal cell, then basad along vein 2A, cutting across axillary cell at alular incision; faint spot of color at base of R_4, sometimes a trace of color at base of second posterior cell; venation as in congeners, with vein at base of second posterior cell sinuous.

Abdominal dorsum pale-yellow pilose and tomentose; incisures of segments two to seven with black hair, black tomentose crossbands on posterior margins of segments two to five, that on segment four and sometimes on segment five reduced to small median spot, tomentum on segments six and seven off-white, lighter than that on rest of abdomen; lateral margin of abdomen with pale-yellow hair basally on segments one and two, then darker-yellow on rest of margin, black hair at posterior corners of segments two to four, hair on apical segments mixed black and yellow. Venter of abdomen pale-yellow pilose and tomentose.

Genitalia with epandrium short, lower apical corner prolonged, prolongation broad, apex bluntly acuminate, lower margin somewhat convex; epiphallus straight, tip curved upward slightly, apex wrinkled, curving evenly to slightly convex lower margin; in dorsal view, epiphallus broad, sides convex, then concave toward base, apex flatly rounded; aedeagus short, pointed downward; basistylus broad, apex pointed; dististylus long, not appreciably hooked at apex.

Female. - Except for denser tomentum and sexual characteristics, as described for male.

Variation. - Face may be entirely black, the femora half-black or nearly all-black; color at base of vein R_4 and at base of second posterior cell, while normally faint, may be wanting on the latter or on both veins. Some females do not have black hair at the side of the second abdominal segment. Hair on posteroventral surface of hind femur may be yellow, mixed yellow and black, or all-black.

Holotype male and allotype. - From 47 mi. SW Chihuahua, Chihuahua, Mexico, IX.6.62 (R. H. and E. M. Painter). Both in U.S. National Museum.

Paratypes. - UNITED STATES: Arizona: Cochise County, 1, Portal, VIII.16.75 (S. I. and S. L. Frommer); Skeleton Canyon, IX.6.58 (H. V. Weems, Jr.); Coconino County, 1, Jacob Lake, VIII.15.58 (W. L. Nutting); Santa Cruz County, 1, pass on E flank Mt. Washington, IX.23.70 (F. S. Bartholomew); 1, Paradise, IX.29.66; Yavapai County, 1, Prescott, VIII.19.53

(P. H. Timberlake). California: Ventura County, 1, 5.5 mi. E Buckthorn, VII.4.68 (B. V. Peterson). Colorado: Huerfano County, 1, 2 mi. W Walsenburg, VIII.24.65 (R. H. and E. M. Painter); Larimer County, 1, Estes Park, VIII.16.55 (R. H. Painter). New Mexico: Lincoln County, 1, 2 mi. SW Capitan, IX.12.62 (R. H. and E. M. Painter); Roosevelt County, 2, 4 mi. SW Elida, IX.13.62 (R. H. and E. M. Painter); Sandoval County, 2, Jemez Springs, Aug. 7; Torrance County, 1, Gran Quivera, VIII.8.65 (H. B. Leech). Utah: Juab County, 2, Dividend, August (T. Spalding); Kane County, 1, jct. Hwy. 14 and 89, VIII.21.57 (R. H. Painter); San Juan County, 1, 5 mi. N Blanding, VIII.27.65 (R. H. and E. M. Painter); 6, 4 mi. S Blanding, VIII.24.67 (J. C. Hall); 3, 22 mi. N Monticello, VIII.30.65 (R. H. and E. M. Painter). CANADA: British Columbia: 1, Richter Pass, 72. mi. W Osoyoos, VI.3.58 (H. and A. Howden). MEXICO: Chihuahua: 26, topotypic, collected with type. Durango: 6, 78 mi. NE Durango, IX.20.59 (R. H. and E. M. Painter); 2, 52 mi. NE Durango, IX.19.59 (R. H. and E. M. Painter). Guerrero: 1, 8 mi. E Taxco, IX.15.63 (R. H. and E. M. Painter). Hidalgo: 13, 6 mi. W Pachuca, VIII.3. 62 (R. H. and E. M. Painter). Nuevo León: 1, 7 mi. W Iturbide, IX.26.63 (R. H. and E. M. Painter). San Luis Potosí: 5, 3 mi. W San Luis Potosí, VIII.10.63 (R. H. and E. M. Painter). Sonora: 1, Yecora, IX.9.59 (W. W. Gibson). Zacatecas: 3, 25 mi. NW Zacatecas, IX.18.59 (R. H. and E. M. Painter); 2, 52 mi. NW Zacatecas, IX.19.59 (R. H. and E. M. Painter).

Paratypes in the collections of the United States National Museum, the University of California at Riverside, and the California Academy of Sciences.

This species may be considered as the northern counterpart of *epheba*, extending from Zacatecas in Mexico to British Columbia in Canada. *P. borea* and *P. epheba* have been taken together in Zacatecas and Durango states. Where they have been found together, they have been collected in about equal numbers.

Paravilla californica Hall, new species

Fig. 6

The yellow pilose abdomen without the black hair, bright fulvous thoracic and scutellar bristles, entirely fulvous legs, and almost entirely hyaline wings serve to distinguish *P. californica* from the congeners.

Male. - Length 8-12 mm. Black; lower half of front, first two antennal segments, face, legs except tips of tarsi, apical abdominal segments fulvous.

Front with black hair on upper half, fulvous hair on lower half; tomentum dull fulvous, paler on lower half of front. Basal antennal segments with black hair; first segment twice as long as second segment, not expanded greatly on inner apical margin; second segment almost twice as wide as long; third segment one and one-half times longer than first two segments combined; base long conical, apical styliform portions equal to conical basal portion. Face with fulvous hair and whitish to pale-yellow tomentum, scattered, short, black hair present in middle above. Proboscis not projecting beyond oral margin. Palpus fulvous, with yellow hair. Side of occiput pale-yellow tomentose, darker-fulvous tomentum toward vertex; cervical fringe pale yellow.

Mesonotum pale-yellow pilose, lighter above base of wing; disk with short black hair, hair across posterior margin shining; tomentum of fulvous scales, well-preserved specimens showing three vague vittae of a little broader and lighter-colored scales, bristles yellow. Pleura pale-yellow to whitish pilose and tomentose. Coxae pale-yellow pilose and tomentose; fore and mid femora with black hair on postero-ventral surface, hind femur without hair, scales on femora and tibiae pale-yellow.

Halter stem testaceous, knob white. Squama testaceous, with fringe of yellow scales. Scutellum with sparse, short, black hair in middle, light-fulvous hair elsewhere; tomentum whitish, fulvous across base and along posterior margin; bristles bright fulvous.

Basicosta with pale-yellow scales and black setulae. Wing hyaline; costal and subcostal cells brownish yellow, as are basal and upper portions of first basal cell; darker, not clearly outlined, spots of color in middle of marginal cell, at base of vein R_{2+3}, on r-m crossvein, and at bases of third and fourth posterior cells, color from spots extending outward along veins for a short distance; vein at base of second posterior cell straight or almost so.

Abdominal dorsum pale-yellow to white pilose; a few short, inconspicuous black hairs in middle on posterior margins of segments two to four, tomentum whitish on basal half of second segment and middle of segments three to seven, otherwise fulvous; lateral margin of abdomen with pale-yellow hair. Venter of abdomen pale-yellow to white pilose and tomentose.

Genitalia with epandrium long and narrow, lower apical corner elongate, extending about as far as median width of elongation, tip bluntly pointed, lower margin concave medially; epiphallus humped somewhat, tip recurved slightly, apex rounded, lower margin straight; in dorsal view, epiphallus rounded broadly at apex, sides concave medially; aedeagus more than half length of epiphallus, curved downward slightly; basistylus long, narrow, not appreciably widened toward base, apex truncate; dististylus short, broad, hooked stongly at apex.

Female. - Except for usual sexual differences of denser tomentum and wider-spaced eyes, the female does not differ from the male.

Variation. - The color in the wings is variable: some specimens have the wings much darker than described, but even in these cases, the second basal and anal cells are hyaline.

Holotype male. - From Sheep Creek Canyon, San Bernardino County, California, V.2.73 (J. C. and E. M. Hall); allotype from Wyman Canyon, White Mtns., Inyo County, California, VII.20.67 (S. and S. Frommer), at light. Both in U.S. National Museum.

Paratypes. - UNITED STATES: Arizona: Pima County, 1, Florida Canyon, Santa Rita Mtns., VIII.22.66 (L. F. La Pré, J. L. Bath). California: Alameda County, 1, Piedmont, VI.

19.12 (Nunenmacher); Kern County, 1, Walker Pass, VI.19.56 (J. C. Hall); Los Angeles County, 3, Los Angeles (D. W. Coquillett); Napa County, 5, Samuel Springs, V.30.53 (J. C. Hall); Riverside County, 19, Riverside, various dates from June to August (J. C. Hall, P. H. Timberlake, G. E. Wallace); San Bernardino County, 1, topotypic, collected with holotype; 1, 28 mi. E Baker, IV.21.64 (P. H. Timberlake); 3, (D. W. Coquillett); San Diego County, 19, 1 mi. E and 5 mi. SE Del Mar, VII.10.63, VII.17.63 (E. I. Schlinger); 1, 3 mi. E Poway, VI.10.65 (J. C. Hall); 1, Rancho Santa Fe, V.11.59 (J. C. Hall); 1, La Jolla, VI.25.63 (E. I. Schlinger); Sacramento County, 2, IX.3.56 (O. Rener); Solano County, 1, Vacaville, V.3.32 (A. J. Morrison); Sonoma County, 2, Monte Rio, 3 mi. S at Dovoul Crk., V.24.70 (T. Plichta); Yolo County, 1, Davis, VIII.13.55 (H. R. Moffitt). Nevada: Nye County, 2, Mercury, Nevada Test Site, VI.11.61. MEXICO: Baja California del Norte: 1, San Vicente, VII.8.63 (J. D. Burchin); Sonora: 1, Kino Bay, VI.15.68 (M. J. Wargo).

Paratypes are located in the collection of the University of California at Riverside and the California Academy of Sciences.

Paravilla castanea (Jaennicke)

Fig. 7

Anthrax castanea Jaennicke, 1867: 338; Aldrich, 1905: 229. *Anthrax hirsuta* Williston, 1901: 279. *Villa (Paravilla) castanea* (Jaenn.) Painter and Painter, 1962: 111, 1965: 438. *Paravilla castanea* (Jaenn.), Hull, 1973: 392.

An uncommon but easily recognized Mexican species. The most striking aspect of this fly is the color pattern of the wings, which resembles that of *Exoprosopa* rather than *Paravilla*; no other species of *Paravilla* is marked with three distinct bands of dark brown in the wings.

According to Painter and Painter (1962) the type is apparently lost. No lectotype is designated here.

Male. - Black; lower half of front, face, first antennal segment, legs except tips of tarsi dark fulvous.

Front with mixed black and fulvous hair, tomentum golden fulvous. Basal antennal segments with mixed fulvous and black hair; first segment twice length of second, widened apically on inner margin; second segment twice as broad as long; third segment almost twice length of two basal segments combined, base elongate conical, apical styliform portion about equal in length to length of broadened basal portion. Face with black hair, fulvous hair next to eye margin and below, tomentum golden fulvous. Proboscis not projecting beyond oral margin. Palpus with black hair. Side of occiput with pale-yellow scales, darker, more fulvous scales at vertex; cervical fringe pale yellow.

Mesonotum with mixed black and fulvous hair on disk, anterior margin entirely fulvous-haired; pale hair with a pinkish or violaceous cast along lateral margin above wing base, mixed with black and fulvous hair on side from wing base to and including humeral callus, tomentum on posterior half of of mesonotum golden yellow, absent on anterior half, bristles fulvous. Pleura fulvous to off-white pilose, lighter color mainly below base of wing, often with a violet cast. Hair on coxae mixed dark-fulvous and black, tomentum absent. Femora with long black hair on posteroventral surfaces, tomentum fulvous with black scales on anterior surfaces; tibiae with fulvous scales and black bristles.

Halter stem brown, knob yellowish. Squama brown with fringe of yellow scales. Scutellum with black hair and golden-yellow scales; bristles fulvous, a few thin, weak, black, bristlelike hairs at sides at base.

Basicosta with fulvous scales and black setulae. Wing veins dark ferruginous, wing infuscated with dark brown; color in three bands - one, at base of wing, extending across basal one-fourth of first and second basal cells, anal and axillary cells; median band extending from vein R_1 posteriorly through basal half of discal cell, apical portions of both basal cells, across base of third and fourth posterior cells, reaching wing margin in anal cell; third band extending from near apex of vein R_1 posteriorly to wing margin in first posterior cell, base of second posterior cell colored, small spots of color at apices of veins R_{2+3}, R_4, M_2, and M_3+Cu_1; first posterior and anal cells narrowed apically; crossvein at base of second posterior cell straight.

Abdominal dorsum fulvous pilose, black hair on segments five to seven and a small spot at sides of segment one, tomentum golden-yellow, scattered, concentrated across posterior margins of segments two to four, triangular black tomentose areas in middle on posterior half of segments two to to five, patch of pure-white scales in middle of segments five and six. Venter of abdomen dark fulvous with a few black hairs, tomentum sparse, pale, concentrated along midline, some specimens with a violet cast to the tomentum.

Genitalia with epandrium a little longer than high, lower apical corner pointed, slightly extended; epiphallus straight or nearly so, apex curved downward, sides at apex flared out, terminating in a downward-projecting sharp point; tip of epiphallus with two dorsal swollen, keel-like ridges, in dorsal view, epiphallus with apex rounded, sides sub-parallel, lower margin of epiphallus swollen, rounded toward apex; aedeagus short, curved downward; basistylus broad, somewhat bluntly pointed on both ends, apex more sharply so; dististylus small, hooked apically.

Female. - Front almost entirely fulvous pilose, and in general, with much more bushy pile and more tomentum, most of which has a violet reflection; otherwise as described for male.

Distribution. - In the Catalog of North American Diptera, *castanea* is listed as occurring in California and Mexico. I have not seen a specimen from California. Specimens examined from Mexico are from: Guerrero: 16 mi. S Chilpancingo, XI. 2.68, 3,100 ft. Hidalgo: Zimapan, X.5.68, 6,200 ft. Mexico: 8. mi. E Texcoco, X.9.68; Chapingo, X.2.68, 7,400 ft. Oaxaca: 3 mi. NW Nachixtlan, IX.10.63, 6,600 ft. Puebla: 19 mi. NE Techuacán, XI.22.68, 7,500 ft.; 3 mi. NE Carmen, Tlaxcala, IX.16.68, 7.500 ft. (all collected by R. H. and E. M. Painter). Jalisco: 10 mi. N Guadalajara, X.16.68 (G. E. Bohart).

Paravilla cinerea (Cole)

Fig. 8

Villa cinerea Cole, 1923: 299. (Type in California Academy of Sciences.). *Villa (Paravilla) cinerea* Cole, Painter and Painter, 1965: 438. *Paravilla cinerea* (Cole), Hull, 1973: 392.

This is a large (11-17 mm. in length) fly distributed throughout the desert regions of the southwestern United States and Mexico. It is similar to and has often been confused with *P. mercedis* (Coq.). Other than by the differences in the male genitalia, *P. cinerea* may be separated from *P. mercedis* by its larger size, infuscated wings, and predominantly black frontal hair.

Male. - Black; two basal antennal segments, face, apical portion of fore femora, all except basal fourth to half of mid and hind femora, tibiae, tarsi except tips ferruginous. Abraded specimens show the posterior margins of the fourth to seventh dorsal abdominal segments narrowly ferruginous; terminal ventral-abdominal segments almost entirely ferruginous.

Eyes separated at narrowest point by twice width of ocellar tubercle. Front with black hair and white to pale-yellow tomentum, scales not reaching ocellar tubercle. First antennal segment more than twice length of second, enlarged slightly on inner apical margin; second segment short, wider than long, both basal segments with stiff black hair; third segment long conical, gradually tapering from base to apex or constricted near middle to form typical styliform portion. Face with stiff black hair and whitish scales. Proboscis short; at most, labellae projecting beyond oral margin. Palpus with black hair. Occiput white pilose and tomentose; cervical fringe white.

Mesonotum whitish to pale-yellow pilose, anterior margin yellow-haired, tomentum white, bristles yellow. Propleuron and anterior portion of mesopleuron with pale-yellowish hair, rest of pleura white or whitish-haired, tomentum whitish. Coxae with yellowish hair, a few black hairs sometimes present apically; tomentum white, long, hairlike. Fore femur with black hair on posteroventral surface, mid femur with pale-yellowish hairs on posteroventral surface, hair absent on hind femur; scales on femora and tibiae pale yellowish, small patch of dark scales at apex of mid femur on anterior surface; bristles black.

Halter stem testaceous, knob creamy white. Squama light testaceous, with fringe of white scales. Scutellum with somewhat appressed yellowish hair, white tomentum, shiny yellow bristles.

Basicosta with whitish scales and black setulae. Wing veins yellowish brown on approximately basal half of wing. Wing hyaline with costal cell yellowish; spots of darker color at bases of first submarginal, discal, and third and fourth posterior cells, and on r-m crossvein. these spots more or less coalesced on more darkly colored specimens and the anal cell showing a light tinge of yellowish brown; first posterior cell narrowed apically, discal cell broad, about four times longer than wide.

Abdominal dorsum pale-yellow to whitish pilose, hair

longest at sides of segments one and two, scattered, individual, shining yellow hairs across posterior margins of segments two to six; tomentum white overall, well-preserved specimens usually with a mid-dorsal spot of brownish tomentum on segments two to five, these spots not normally evident. Venter of abdomen white pilose and tomentose.

Genitalia with epandrium nearly square, lower apical corner prolonged and bluntly pointed; epiphallus thick, straight, curved upward slightly apically, lower margin rounded gently, not swollen; in dorsal view, epiphallus broad, tip bluntly rounded, sides convex beyond middle; aedeagus less than half length of epiphallus, curved downward slightly; basistylus broad, base rounded, apex somewhat square; dististylus thick, hooked apically.

Female. - Except for denser pile and tomentum, the female does not differ appreciably from the male.

Variation. - The middle of the face may or may not be dark; the legs vary from as described to entirely fulvous except for the base of the fore femora; some specimens, especially females, have testaceous scales scattered over the thoracic dorsum; wing coloration may be light or dark, and in the darkly colored wing the color is more extensive.

Distribution. - Cole (1923) described this species from specimens collected at Angeles Bay in Baja California in May. The distribution in the Catalog of Diptera of North America is given as Baja California, Mexico, Arizona, and Texas. Specimens examined by me are from the deserts of California, Arizona, and northern Mexico.

These flies are on the wing early in the year, since specimens were taken from March to June. They feed mainly on composites, e.g., *Encelia farinosa* and *Senecio* spp.

Paravilla consul (Osten Sacken)

Fig. 9

Anthrax consul Osten Sacken, 1886: 125; Coquillett, 1894: 98; Williston, 1905: 229. (Type in British Museum, Natural History collection.). *Villa (Paravilla) consul* (O. S.), Painter 1933: 10; Painter and Painter, 1962: 100, 1965: 438. *Paravilla consul* (O. S.), Hull, 1973: 392.

This species, and with *P. extremitis* (Coq.), *P. diagonalis* (Lw.), *P. nigriventris*, and *P. eminens* form a group of closely related species characterized primarily by the venter of the body being entirely black pilose. Within this group, *P. consul* is the most readily recognized and by far the most common. It may be separated by the following combination of characters: thoracic pleural area being entirely black pilose, medially interrupted pale bands of tomentum on the second, third, and fourth dorsal abdominal segments, and rounded apical corner of the epandrium.

Male. - Black; knob of halter testaceous to creamy white.

Eyes separated at narrowest point by twice width of ocellar tubercle. Front with black hair and golden scales, black scales in middle on lower half, upper half of front without scales. First antennal segment twice length of second segment, not appreciably expanded apically; second segment a little wider than long, both basal segments with black hair; third segment long conical at base, apical styliform portion longer than conical basal portion, entire segment nearly twice length of two basal segments combined. Face with stiff black hair, tuft of black hair at tip of oral margin, scales in middle above black, contiguous with those on front, a few thin, scattered, golden scales laterally. Palpus with black hair. Occiput with yellow scales and hair, a few black hairs present at vertex; cervical fringe of black hair below, yellow above.

Mesonotum with black hair on disk and across posterior margin, longer posteriorly, collar and sides above root of wings with pale-yellow hair, tomentum yellow, appressed anteriorly, then gradually becoming longer and semi-erect posteriorly, bristles black or pale. Upper third of thoracic pleural area yellowish-white pilose, lower half black pilose, black hairlike tomentum on sternopleuron. Coxae and femora with black hair and scales.

Squama white, with fringe of white scales. Scutellum with black hair and bristles, tomentum yellow, long, hairlike.

Basicosta with black scales and setulae. Wing infuscated on basal half with dark brown, apical half or more hyaline; color runs diagonally in step fashion basad from tip of subcosta to extreme base of axillary cell, passing through base of first submarginal cell, beyond r-m crossvein in first posterior cell to r-m crossvein in discal cell, across bases of third and fourth posterior cells, through anal cell a little beyond posterior crossvein, margin of color more or less sharply defined, both basal cells entirely colored; discal cell three times longer than greatest width; vein at base of second posterior cell bent slightly.

Abdominal dorsum white to pale-yellow pilose on segments one and two, short on dorsum, longer and bushy laterally, rest of abdomen with black hair; tomentum pale yellowish across bases of segments two to four, bands emarginate or separated medially, segments six and seven with pure white to silver-white scales, rest of abdomen black tomentose; lateral margin of abdomen pale-yellow to white pilose on segment one and basal half of segment two, rest of sides with black hair. Venter of abdomen entirely black pilose and tomentose.

Genitalia with epandrium nearly as wide as long except for basal pedicle, apical lower corner rounded broadly; epiphallus bent slightly downward, tip recurved, acuminate; apex of epiphallus blunt, wrinkled, lower margin broadly rounded in middle; in dorsal view, tip of epiphallus convex slightly, sides subparallel, divergent basally; aedeagus half length of epiphallus, curved downward; basistylus broad, apex rounded; dististylus hooked apically.

Female. - Except for genitalia and usual sexual differences of more widely separated eyes and denser tomentum, the female does not differ from the male.

Variation. - Two distinct phenotypes occur: (1) Specimens with the wings marked as described, and (2) specimens in which the wing color is not sharply outlined, with the color often reaching the base of vein R_4 and often with a spot of color at the base of the second posterior cell. This second phenotype has been found in several populations together with the more common form. The male genitalia do not differ. In both forms the thoracic bristles may be entirely black, entirely pale, or mixed black and pale.

Distribution. - According to the Catalog of Diptera of North America, locations are Guatemala and Arizona. Williston (1901, p. 280) lists several localities in Mexico. *P. consul* is primarily a Mexican species, found from Costa Rica in the south through Mexico to southern Arizona and southern New Mexico. The specimen from North Dakota mentioned by Coquillett (1894, p. 98) as possibly belonging to this species is actually *P. separata* (Walker).

P. consul has been collected in August and September.

Paravilla cunicula (Osten Sacken)

Fig. 10

Anthrax cuniculus Osten Sacken, 1886: 125. (Type in British Museum, Natural History collection.). *Villa (Paravilla) cuniculus* (O. S.), Painter, 1933: 10; Painter and Painter, 1962: 101, 1965: 439. *Paravilla cunicula* (O. S.), Hull, 1973: 392.

One of the smaller (8-11 mm. in length) species, resembling *P. diagonalis* (Loew). It may be separated by the following characters: lower half of the body with white tomentum; base of abdomen white pilose; wing with basal half infuscated, apical half hyaline.

Male. - Black; first antennal segment, face fulvous, legs dusky to partially fulvous.

Eyes separated at narrowest point by one and one-half times width of ocellar tubercle. Front with black hair and pale testaceous scales. First antennal segment twice length of second, expanded apically on inner margin; second segment twice as wide as long, both basal segments with black hair; third segment short conical at base, much broader than second segment, apical styliform portion a little shorter than basal portion. Face with black hair and pale-testaceous scales. At most, tips of labellae projecting beyond tip of oral margin. Palpus testaceous to black, with yellowish hair. Occiput white tomentose, with pale-yellow hair on vertex; cervical fringe white.

Mesonotum with short black hair on disk, longer posteriorly, collar of pale-yellow hair, white hair along lateral margin, tomentum absent except across posterior margin and on post-alar declivity, where it is white; bristles yellow. Pleura whitish to very pale-yellow pilose, darker yellow on propleuron, tomentum white. Coxae with mixed yellow and black hair and testaceous scales; hair on femora brown, scales whitish, white scales on tibiae.

Halter stem yellowish, knob white. Squama pale testaceous, with white fringe. Scutellum with black hair and bristles, tomentum white.

Basicosta with testaceous scales and black setulae. Wing infuscated on basal half, apical half hyaline; color runs from tip of subcosta diagonally basad to near apex of anal cell, base of discal cell colored to r-m crossvein, both basal cells colored, axillary cell hyaline, color along vein A_2 only; vein at base of second posterior cell straight; r-m crossvein before middle of discal cell.

Abdominal dorsum white pilose and tomentose, black hair across posterior margins of segments three to five, black or dark-brown scales in middle of posterior margins of segments two, three, and five, segment four nearly entirely dark-brown to black tomentose; lateral margin of abdomen white pilose. Venter of abdomen with white hair on basal half, dark hair on apical half, tomentum dark testaceous, with white spots of scales in middle on posterior margins of segments two to six.

Genital capsule white-haired; epandrium narrow, lower margin concave, short, apical margin slanted basad, lower apical corner rounded; epiphallus short, curved downward slightly, apex curved upward, broad, lower margin with slight bulge medially, apex wrinkled; in dorsal view, epiphallus rounded apically, sides concave below middle, then diverging toward base; aedeagus a little more than half length of epiphallus, curved downward, apex acuminate; basistylus short, broad, apex bluntly pointed; dististylus long, narrow, apex with a small hook.

Female. - Mesonotum white tomentose; abdominal dorsum white pilose at base, with black hair on posterior half of all segments except the first; tomentum white, black tomentose crossbands on posterior portions of segments two and three, segment four almost all black tomentose, dark-brown scales at apex of segment five and base of segment six; lateral margin of abdomen with black hair from posterior corner of second segment to apex. Otherwise as described for male.

Variation. - In the male, a spot of dark scales on the second segment may be present or absent; the dark abdominal scales may be black, dark brown, or a mixture of the two.

In some, especially males, the abdomen is nearly entirely white pilose and tomentose. White tomentum may be present on the mesonotum of males, but is never as dense as in females. Leg coloring varies from entirely dull fulvous to nearly entirely dusky, but is never black. Specimens from the northern end of the distribution range tend to be darker.

Distribution. - The Catalog of Diptera of North America lists the distribution as Mexico and Arizona. This is extended to Texas and California. The species is not common in collections and is usually collected from April to June. A few specimens have been taken in California in September.

Paravilla deserta Hall, new species

Fig. 11

Much like *P. acutula*, but separable by the black or dusky fore femora, infuscated second basal cell, evenly tapered third antennal segment, with apical styliform portion usually shorter than long conical basal portion, and usual absence of black tomentose crossbands on the abdominal dorsum. In some specimens, especially females, there may be an enlarged median spot of black tomentum on the second abdominal segment. In the male, the lower apical corner of the epandrium is prolonged, narrow, and rounded apically.

Male. - Length 8-12mm. Black; lower portion of front, first antennal segment and sometimes second, face fulvous, legs except tarsi ferruginous, femora dusky.

Front and face black pilose and pale-yellow tomentose. Antenna with black hair; first segment short, about one and one-half times longer than second segment, cylindrical, not enlarged on inner apical margin; second segment as long as wide; third segment about twice as long as first two segments combined, base long conical, apical styliform portion much shorter than swollen basal portion. At most, labellae projecting beyond tip of oral margin. Palpus with black hair. Occiput whitish pilose and tomentose, with a few yellow hairs present at vertex; cervical fringe yellowish white.

Mesonotum black pilose on disk, collar and side anterior to wing base pale-yellow pilose, long yellow hairs across posterior margin; tomentum flavotestaceous, white along lateral and across posterior margins, tomentum dense along side, scattered along posterior margin, bristles yellow. Pleura pale-yellowish pilose and tomentose. Coxae with pale-yellow and black hair and pale-yellow scales, femora with black hair and yellowish scales, anterior surface of fore femur almost entirely covered with black scales; tibiae with yellowish scales, setulae and bristles black.

Halter stem yellow, knob white. Scutellum with black hair, white at sides near base, tomentum white, becoming yellowish toward posterior margin; bristles yellow.

Basicosta with mixed flavotestaceous and black scales and black setulae. Wing hyaline, basal half colored with light brown; color extends diagonally basad in an irregular line from the tip of subcosta to middle of axillary cell, base of discal cell colored to r-m crossvein, both basal cells entirely colored, anal cell colored to a little beyond posterior crossvein; color not reaching margin in axillary cell; slightly darker spots of color at base of vein R_{2+3}, on r-m crossvein, and at bases of third and fourth posterior cell; vein at base of second posterior cell slightly sinuous.

Abdominal dorsum white pilose, scattered yellow hair present on segments two to five, black hair across posterior margins of segments two to five, not reaching lateral margin of abdomen; white tomentum on basal two-thirds of second segment, at sides at base of third segment and on all of segments six and seven, flavotestaceous scales elsewhere; most specimens with small median spots of dark-brown scales on segments two and three; lateral margin of abdomen with white hair, becoming yellowish apically. Venter of abdomen white pilose and tomentose.

Genitalia with epandrium short, not much longer than high, lower apical corner prolonged greatly, at least four times its median width, lower margin concave slightly; epiphallus almost straight, broad, tip not appreciably curved upward, apex rounded, lower margin convex; in dorsal view, epiphallus broad, apex rounded, sides convex; aedeagus shorter than half length of epiphallus, curved downward; basistylus somewhat narrowed, base and apex bluntly pointed; dististylus long, hooked slightly at apex.

Female. - Except for denser tomentum, more black hair on abdominal dorsum, and black scales on anterior surface of mid femur, does not differ appreciably from male.

Variation. - Face and legs may be ferruginous to black; all femora may be black or dusky, or only the fore and mid femora may be darkened; black scales may be present on all femora, or be as described for both sexes; second basal cell may be subhyaline; dark-brown tomentose spots on abdominal segments two and three may be absent, very small, or quite pronounced, and/or on segment four as well.

Holotype male and allotype. - From 5 mi. S Ray, Pinal County, Arizona, IV.5.61 (R. H. and E. M. Painter). Both in U.S. National Museum.

Paratypes. - UNITED STATES: Arizona: Cochise County, 4, Southwestern Research Station, IX.11.66 (P. H. Arnaud, Jr.); Coconino County, 1, The Gap, VIII.24.57 (R. H. and E. M. Painter); Maricopa County, 1, 10 mi. S Gila Bend, IV.13.61 (R. H. and E. M. Painter); 1, 18 mi. S Gila Bend, IV.18.65 (M. A. Cazier, J. M. Davidson, S. A. Gorodenski); Pima County, 9, Organ Pipe Cactus Natl. Monument, IV.14.61 (R. H. and E. M. Painter); Pinal County, 11, topotypic, collected with types; Yuma County, 8, 11 mi. W Welton, IV.12.61 (R. H. and E. M. Painter); 1, Welton, IV.15.38 (F. H. Parker); 2, Yuma, X.20.58 (T. R. Haig). California: Riverside County, 2, Corn Springs, Chuckwalla Mtns., III.28.67 (J. C. Hall); 1, Andreas Canyon, IV.11.36 (P. H. Timberlake); San Bernardino County, 1, 7 mi. SW Cima, V.1.66 (P. H. Timberlake). Colorado: Rio Grande County, 1, 7. mi. SE Del Norte, VIII.25.63 (R. H. and E. M. Painter). Montana: Granite County, 1, VIII.11.31 (R. H. Beamer). Nevada: Nye County, 4, Mercury, Nevada Test Site, V.28, VI.10, VI.11.61; White County, 2, Baker, VIII.21.57 (R. H. and E. M. Painter). Utah: Carbon County, 1, 7 mi. SE Dragerton, VIII.25.65 (R. H. and E. M.

Painter); Garfield County, 1, Hatch, IV.25.57 (R. H. and E. M. Painter); San Juan County, 1, 22 mi. N Monticello, VIII.31.65 (R. H. and E. M. Painter). Wyoming: Niobrara County, 4, Lance Creek, VIII.15.95 (W. M. Wheeler). MEXICO: Chihuahua: 2, Santa Barbara, IX.5.62 (R. H. and E. M. Painter), Coahuilla: 1, 13 mi. NE Torreón, IX.21.59 (R. H. and E. M. Painter), Nuevo León: 1, Linares, IV.25.66 (R. H. and E. M. Painter), Sinaloa: 1, 2 mi. S Ocurahuí, X.1.70 (D. E. Breedlove), Sonora: 8, 75 mi. S Hermosillo, IV.25.61 (R. H. and E. M. Painter); 3, 4 mi. S and 7 mi. S Magdalena, IV.19.61 (R. H. Painter).

Paratypes in the collections of the United States National Museum, the University of Arizona, the University of California at Riverside, and the California Academy of Sciences.

Paravilla diagonalis (Loew)

Fig. 12

Anthrax diagonalis Loew, 1869: 21; Coquillett, 1894: 97. (Type in Harvard Museum of Comparative Zoology; see below.). *Villa (Paravilla) diagonalis* (Lw.), Painter, 1933: 10; Painter and Painter, 1965: 439. *Paravilla diagonalis* (Lw.), Hull, 1973: 392.

Similar to and often misidentified as *P. consul*. The two are easily separated not only by the male genitalia, but also by the white-haired thoracic pleural area and more northern distribution of *P. diagonalis*.

Male. - Length 7-12 mm. Black; genital capsule brownish.

Front and face with black hair and yellowish tomentum, lower portion of front above antennae devoid of hair in middle, scales not reaching to ocellar tubercle; scales on face frequently translucent, reflective, showing whitish to yellowish depending on light source. Tip of oral margin with tuft of overhanging, stiff, black hair. First antennal segment cylindrical, expanded slightly on inner margin at apex, twice as long as second segment; second segment a little wider than long, both basal segments with black hair; third segment one-third longer than two basal segments combined, base elongate conical, constricted, but not strongly so, before middle, then tapering to styliform apical portion. At most, labellae projecting beyond tip of oral margin (several specimens have the proboscis projecting well beyond margin, as in *Rhynchanthrax*). Palpus with black hair. Occiput with pale-yellowish scales; cervical fringe pale-yellowish above, black below.

Mesonotum with short black hair on disk, longer toward scutellum, collar and lateral margin with pale-yellow hair, tomentum yellowish, covering mesonotum uniformly, bristles pale-yellow; one or two black bristles may be present on post-alar callus. Pleura pale-yellow pilose and tomentose, lower half of propleuron with black hair. Legs with black hair and scales.

Halter stem testaceous, with small black scales on posterior surface, knob whitish. Squama pale, with fringe of pale-yellowish scales. Scutellum with short, scattered, black hair, yellowish tomentum overall; bristles black.

Basicosta with black scales and black setulae, a few yellowish scales along lower margin at base. Wing infuscated with a light brownish-yellow on basal half or less;

color runs from tip of subcostal vein diagonally basad to basal fourth of axillary cell, color in discal cell not reaching r-m crossvein, both basal cells infuscated, basal portion of second basal cell lighter in color, color in anal cell not extending beyond posterior crossvein, basal fourth of axillary cell infuscated along vein 2A only; r-m crossvein before middle of discal cell; crossvein at base of second posterior cell more or less sinuous.

Abdominal dorsum virtually without hair, that present short, black, and inconspicuous, overlying black tomentum and in middle of posterior margin of first segment; crossbands of pale-yellow tomentum on bases of segments two to four, segments six and seven entirely pale-yellow tomentose, apical halves of segments two to four with black tomentum, fifth segment entirely black tomentose; lateral margin of abdomen with pale-yellow to whitish hair on segment one and nearly all of segment two, rest of margin black-haired. Venter of abdomen black pilose and tomentose, long whitish hair on segments one and two and sometimes on three.

Genitalia with epandrium moderately narrow, lower apical corner jutting, rounded, lower margin concave before middle; epiphallus short, thick, arched slightly, tip not appreciably curved upward, apex truncate, wrinkled on upper half, lower margin swollen apically; in dorsal view, epiphallus flat, apex rounded, sides convex, converging below middle; aedeagus short, broad, curved downward; basistylus short, broad, apex bluntly pointed; dististylus hooked apically.

Female. - Pile and tomentum comparatively denser, black

hair on abdominal dorsum more conspicuous, not otherwise different from male.

Holotype female. - Almost completely denuded, but enough traces of hair and scales are left to permit comparisons with better-preserved individuals. When denuded, *P. diagonalis* and *P. consul* are difficult to separate. In the original description, Loew stated that the hair and tomentum of the thorax and scutellum were black, appearing yellow. The few scales left on the type are all yellow.

Distribution. - In the Diptera Catalog, the distribution is listed as California, Texas, and Kansas. I have examined specimens not only from California and Texas, but also from Arizona, New Mexico, and Mexico. Judging from the several hundred specimens seen, *P. diagonalis* appears to be a desert species, making the record from Kansas doubtful. The distribution of *P. diagonalis* is more northern than that of *P. consul*: specimens of *P. diagonalis* have not been encountered south of Chihuahua or Sonora in Mexico, while *P. consul* has been found as far south as Guatamala City, Guatamala.

Paravilla edititoides (Painter)

Fig. 13

Villa (Paravilla) edititoides Painter, 1933: 11; Painter and Painter, 1965: 439. (Type in U.S. National Museum.). *Paravilla edititoides* (Paint.), Hull, 1973: 392. *Villa (Paravilla) fulvicoma* (Coq.), Painter, 1933: 11 (in part, misidentification).

In habitus, *P. edititoides* is very similar to *P. fulvicoma* (Coq.). In fact, the two are so much alike that the only reliable means of separating them is by the male genitalia. In *P. edititoides* the aedeagus is short and curved downward, as in most other *Paravilla*, while in *P. fulvicoma* the aedeagus is straight and almost as long in extent as the epiphallus.

The three species *P. edititoides*, *P. fulvicoma*, and *P. palliata* form a complex of species which, for the most part, overlap in their external characteristics. Of the three, *P. palliata* is the easiest to separate because of the black bristlelike hair between the wing bases and the humeral calli.

Male. - 10-15 mm in length. Black; first antennal segment, face, legs fulvous.

Front and face with black hair and golden-yellow tomentum; two basal antennal segments with black hair, first antennal segment short, one and one-half times longer than second segment, expanded slightly apically; second segment as long and wide; third segment a little less than twice length of first two segments combined, base long conical, apical styliform portion longer than swollen basal portion. Proboscis not projecting. Palpus with black hair. Occiput pale-yellow tomentose, with yellow hairs at vertex; cervical fringe pale yellow.

Mesonotum yellow pilose and tomentose, tomentum scattered, denser on posterior half of notum, patch of whitish hair above root of wing; bristles pale. Pleura pale-yellow to whitish pilose and tomentose. Coxae with black and yellow hair, tomentum yellow; hair on posteroventral surface of fore femur black, that on mid and hind femora yellow, scales on femora and tibiae pale yellow.

Halter stem fulvous, knob white. Squama testaceous, with fringe of pale-yellow scales. Scutellum yellow pilose and tomentose; bristles yellow.

Basicosta with yellow scales and black setulae. Basal half of wing infuscated with brown, apical half hyaline; color runs from apex of subcosta diagonally basad to apical third of anal and axillary cells, both basal cells infuscated, color in discal cell not reaching r-m crossvein except along medial vein, color not reaching posterior margin of wing in axillary cell; r-m crossvein at middle of discal cell; crossvein at apex of discal cell straight or nearly so.

Abdominal dorsum yellow pilose and tomentose, that on segments six and seven whitish, a few conspicuous black hairs on posterior margins of segments three to five; lateral margin of abdomen yellow pilose. Venter of abdomen pale-yellow pilose and whitish tomentose.

Genitalia with epandrium short, nearly square, apical lower corner rounded, lower margin straight; epiphallus arched slightly, tip recurved, aped rounded, wrinkled, lower margin bulbous at apex; in dorsal view, epiphallus with apex rounded, sides subparallel; aedeagus short, half length of epiphallus, bent downward; basistylus rounded at base, apex bluntly pointed; dististylus short, not strongly hooked at apex.

Female. - Like the male except for sexual characteristics.

Variation. - The description above was drawn from paratopotypic specimens from LaVeta, Colorado. This species is variable in respect to the coloring of the antennae, face, and legs, color of the pile, and presence of black hair on the abdomen. Some specimens have the antennae and face entirely black, and in others the femora vary from as described through dusky to nearly all-black. The hair on the front may be all black, all fulvous, yellow, or a mixture of black and fulvous. The male terminalia vary somewhat in degree of sclerotization along the margins of the various parts; these differences have been noted in individuals from the same population. Some specimens, especially those from Mexico, have small triangles of black tomentum in the middle of the posterior halves of segments two to four.

Distribution. - On the basis of the specimens studied, *P. edititoides* is found ranging from Saskatchewan, Canada, through eastern Nebraska south to Jalisco, Mexico. It is a dominant species of *Paravilla* throughout the lower elevations of the Rocky Mountains from Wyoming to New Mexico. It is on the wing primarily during the late summer and early fall (August-October).

Some of the specimens from Painter's type series do not belong here. The specimens he listed from California, Oregon, and the following localities in Texas belong with *P. fulvicoma*: Kingsville, Sweetwater, and Brownwood. The series of specimens reared by Painter from *Diadasia afflicta* Cresson at Manhattan, Kansas, differs slightly from the forms from the Rocky Mountains. It is left here for the present, pending additional material.

Paravilla eminens Hall, new species

Fig. 14

This member of the consul group differs from the others by the presence of pale-colored hair on the basal half of the lateral margin of the third abdominal segment, presence of pale scales at the sides of the fifth abdominal segment, and projecting lower apical corner of the epandrium.

Male. - Large size, 12-15 mm. in length. Black; knobs of halteres white.

Front, antennae, and face with black hair, front and face with golden-yellow scales, black scales in middle above on face extending upward between antennae to lower fourth of front. First antennal segment cylindrical, twice length of second segment, not expanded apically; second segment short, as wide as long; third segment short conical at base; styliform portion twice length of basal portion. At most, tips of labellae projecting beyond oral margin. Palpus with black hair. Occiput yellow pilose and tomentose, scattered black hair on upper half; cervical fringe with yellow hair on upper half, black hair on lower half.

Mesonotum with short black hair on disk, longer toward posterior margin, pale-yellow hair on anterior, lateral, and posterior margins, tomentum yellow, scattered; bristles pale yellow. Pleura, except for pale-yellow hair on upper third, entirely black pilose and tomentose. Coxae and legs with black hair and scales.

Squama white, with fringe of white scales. Scutellum

with black hair and bristles; tomentum yellow longer at sides at base.

Basicosta with black scales and setulae. Wing infuscated on basal half, apical half or more hyaline; color runs diagonally basad from tip of subcostal vein through base of discal cell as far as r-m crossvein, across bases of third and fourth posterior cells, on all but about apical third of anal cell into middle of basal half of axillary cell, both basal cells entirely infuscated; wing otherwise as in *consul*.

Abdominal dorsum whitish pilose on sides of segments one to four, rest with black hair; crossbands of pale-yellow tomentum on basal halves of segments two to four, bands narrowed medially and often interrupted, patch of pale-yellow scales on sides of fifth segment, segments six and seven with silver-white scales, black tomentum elsewhere; lateral margin of abdomen with pale-yellow to whitish hair on segments one to four, rest with black hair. Venter of abdomen entirely black pilose and tomentose.

Genitalia with epandrium narrow, lower margin deeply concave, lower apical corner prolonged for a distance about seven times width at middle; epiphallus thick, apex not curved upward, truncate, wrinkled, lower margin straight; in dorsal view, epiphallus flat, rounded, sides concave slightly near middle; basistylus short, basal half thickened, apex bluntly pointed; dististylus short, thickset, hooked apically.

Female. - Does not differ appreciably from male.

Holotype male and allotype. - From 3 mi. N Chilpancingo, Guerrero, Mexico, VIII.31.59 (R. H. and E. M. Painter). Both in U.S. National Museum.

Paratypes. - MEXICO: All but one collected by R. H. and E. M. Painter. Guerrero: 2, topotypic, collected with type; 1, 24 mi. N Chilpancingo, VIII.31.59; Morelos: 1, 3 mi. E Cuernavaca, VIII.10.62; 1, 35 mi. S Cuernavaca, VIII.27.59; 1, Cuernavaca, VIII.7.38 (L. J. L.).

Paravilla emulata (Painter)

Fig. 15

Villa (Paravilla) emulata Painter, 1962: 103; Painter and Painter, 1965: 439. (Type in U.S. National Museum.). *Paravilla emulata* (Paint.), Hull, 1973: 392.

Painter and Painter (1962) redescribed Bigot's badly denuded type of *P. eurhinata*, the only known specimen. Comparison of specimens of *P. emulata* with Painter's redescription of *P. eurhinata* shows that the two species are similar; they may be separated by the absence of tufts of black hair on the apices of the abdominal segments along the lateral margins in *P. emulata*.

Those paratypes of *P. emulata* from California listed by Painter represent an undescribed species, described herein as *P. montivaga*. *P. emulata* may be recognized by the intense coloring in the wing, especially at the bases of the second submarginal and second posterior cells, and by yellow hair on the posteroventral surface of the mid and hind femora.

Painter's description is, for the most part, complete

and available. The following remarks can be added to it: in none of the specimens studied are there black hairs on the scutellum; not all specimens have black hair present on the abdomen, nor is the hair above the root of the wings white - it is, rather, a very pale yellow.

The epandrium is long and narrow, nearly twice as long as wide, the lower apical corner bluntly pointed, lower margin more or less straight; epiphallus straight, tip recurved slightly, apex rounded with a small median depression, lower margin straight; in dorsal view, epiphallus with apex rounded, sides concave; aedeagus short, straight, pointed downward; basistylus long, rounded basally, apex bluntly rounded; dististylus long, deeply hooked at apex.

Distribution. - In addition to Painter's type series, specimens have been studied from New Mexico, Arizona, and Colorado. All were collected in August and September.

Paravilla epheba (Osten Sacken)

Fig. 16

Anthrax ephebus Osten Sacken, 1886: 124. (Type in British Museum, Natural History collection.). *Villa (Paravilla) epheba* (O. S.), Painter and Painter, 1962: 102, 1965: 439. *Paravilla epheba* (O. S.), Hull, 1973: 392.

This Mexican species that ranges from southern Arizona to Honduras is another that has a spot of color at the bases of the second submarginal and second posterior cells. *P. epheba* is possibly most closely related to its more northernly counterpart *P. borea*. The most obvious differences between the two lie in the darker coloring in the wing, especially at the bases of the second submarginal and second posterior cells in *P. epheba*, and in the male genitalia. In *P. epheba*, the lower apical corner of the epandrium has a broad and apically rounded projection; in *P. borea*, the projection of the epandrium is pointed apically.

The original description, while adequate, does not clearly elucidate all of the salient characteristics. In conjuction with Osten Sacken's description, the following should serve to define this species.

Male. - Ground-color black; somewhat ferruginous on side of face, legs fulvous, basal half or more of femora and tips of tarsi black.

Hair pale yellow, white hair along lateral margin of mesonotum above root of wing, on posterior corner of mesopleuron, and at sides of abdomen at base; black hair on front, antennae, face, coxae, and femora, middle of mesonotum, disk of scutellum, and abdominal segments four to six. Tomentum yellowish, darker on front and face.

Wing color reaches base of vein R_4 in first submarginal cell.

Abdomen banded with black, pale-yellowish with black tomentum across posterior margins of segments two to five, most extensive on two, then reduced in amount posteriorly; scales on six and seven whitish. Venter of abdomen whitish to testaceous pilose and tomentose.

Genitalia with epandrium narrow except for base, lower apical corner projecting, rounded apically, lower margin concave; epiphallus humped, tip curved upward slightly, apex rounded, lower margin straight or nearly so; in dorsal view, epiphallus broad, apex rounded, sides subparallel; aedeagus short, pointed downward; basistylus broad, base round, apex bluntly pointed; dististylus short, hooked apically.

Female. - Hair and tomentum usually darker yellow. Tomentum denser; otherwise like male.

Variation. - The pile may be nearly all white or quite dark yellow. The bands of black tomentum on the dorsum of the abdomen vary from complete, that is, reaching the lateral margins of the abdomen, to short. The black tomentum on segments three to five may be confined to the posterior margins of the segments or may extend medially to the anterior margins. The color in the first submarginal cell of the wing may be dark or light, sometimes reaching base of R_4 along vein R_5 only. The hind femur and coxae may have yellow or black hair. The toothlike projections at the base of the claws mentioned by Osten Sacken are rudimentary pulvilli. Painter and Painter (1962, p. 105), in their key to the species with a spot of color on vein R_4, key *epheba* as having the first antennal segment, tibiae, and tips of femora yellowish or brownish; the majority of the specimens I have seen have black antennae.

Distribution. - This species is found predominantly in Mexico, from Chihuahua to Oaxaca, extending south to Honduras. A few specimens have been taken in Arizona in late summer.

Paravilla eurhinata (Bigot), new combination

Anthrax eurhinatus Bigot, 1892: 355. (Type in British Museum, Natural History collection). *Villa (Paravilla) eurhinata* (Big.), Painter and Painter, 1962: 102, 1965: 439. *Villa eurhinata* (Big.), Hull, 1973: 392.

The type female is apparently a unique. I know of no other specimens that might possibly be confused with *P. eurhinata*.

Painter and Painter (1962) redescribed the type, and my examination of Bigot's specimen shows the redescription to be accurate. The following diagnosis, together with the characters given in the key, distinquishes this species: first antennal segment, sides of face, legs fulvous, rest black; pile black on front and face, pale yellow on thorax and abdomen; tomentum on front and face dull fulvous; hair on posteroventral surface of fore and mid femora black; propleuron with black and pale-yellow hair; abdomen with black tomentum across posterior margins of segments two to four (apparently so, since the type has the middle of the abdominal dorsum abraded), lateral margin of abdomen with black hair from segment two to apex.

Distribution. - Known only from Mexico, although I have a single specimen from Honduras that I tentatively place with *P. eurhinata*.

Paravilla extremitis (Coquillett)

Fig. 17

Anthrax extremitis Coquillett, 1902: 138. (Type in U.S. National Museum.). *Villa (Paravilla) extremitis* (Coq.), Painter and Painter, 1965: 439. *Paravilla extremitis* (Coq.), Hull, 1973: 392.

Similar to and often confused with *P. consul*, but actually more like *P. eminens*. *P. extremitis* may be separated from *P. consul* primarily by the absence of pale tomentose crossbands on the abdominal dorsum, absence of black hair on the lateral margin of the abdomen anterior to the fourth segment, and the somewhat elongated lower apical corner of the epandrium.

Coquillett based his original description on 15 specimens. Only the holotype remains in the U.S. National Museum.

Male. - Black; oral margin may be tawny, knob of halter creamy white.

Eyes separated at narrowest point by twice width of ocellar tubercle. Front with black hair and scattered golden scales, a few black scales present medially. Antenna with black hair, first segment one and one-half times longer than second segment, not enlarged apically; second segment almost as long as wide; third segment elongate conical at base, styliform portion one and one-half times longer than basal portion, third segment one and one-half times longer than two basal segments combined. Face with black hair, black scales in middle above, sides with golden-yellow scales. Palpus with black hair. Proboscis not projecting. Side of occiput with pale-yellow scales, short black hair near vertex; cervical fringe yellowish above, black or dark brown below.

Mesonotum black-haired, longer toward scutellum, yellow hair on anterior and lateral margins, pile a little lighter-yellow above root of wing, tomentum of scattered, golden yellow, semi-erect, short hair, bristles yellow. Upper half of pleura with pale-yellow hair, lower half with black hair, tomentum black. Coxae and femora with black hair and scales, tibiae with black scales.

Squama whitish, with fringe of whitish scales. Scutellum with black hair, tomentum golden-yellow; pale yellow bristles.

Basicosta with black scales and setulae. Wing infuscated with dark brown on upper basal portion; color extends from tip of subcosta diagonally basad to middle of anal cell, both basal cells infuscated, color in discal cell not reaching r-m crossvein except along vein M_1, axillary cell hyaline except along vein 2A; discal cell broadened apically; crossvein at base of second posterior cell sinuous.

Abdominal dorsum pale-yellow pilose, with black hair on posterior margins of segments four to six, white hair on seventh segment, black tomentum on segments two to five, silver-white tomentum on segments six and seven, bases of segments two to four with short bands of yellow tomentum at sides, bands not reaching middle of abdomen; lateral margin of abdomen with pale-yellow hair on segments one to four, black on five, white on segments six and seven. Venter of abdomen black pilose and tomentose.

Genitalia with epandrium narrow, lower margin concave, lower apical corner projecting for a distance more than twice its width; epiphallus humped, tip not curved upward, apex blunt, wrinkled, lower margin straight or almost so;

in dorsal view, epiphallus with apex rounded, sides subparallel, concave toward base; aedeagus more than half length of epiphallus, curved downward slightly; basistylus broad basally, apex bluntly pointed; dististylus short, not deeply hooked at apex.

Female. - As described for male, except for more abundant tomentum, and silver-white scales at apex of abdomen not reaching lateral margins.

Variation. - A few specimens have been seen that have pale tomentose stripes at the sides of the abdomen complete, that is, extending across the dorsum, although the bands are very narrow medially. A few females lack silver-white tomentum at the tip of the abdomen.

Distribution. - In the Catalog of Diptera of North America, the distribution is given as Mexico and Colorado. I have additional specimens from New Mexico. Specimens studied were collected in July and August from the following states in Mexico: Morelos, Guerrero, Oaxaca, Distrito Federal, Mexico, Puebla, Durango, Aguascalientes, and Chihuahua, and from Colorado: Huerfano County in August and September; and New Mexico: Valencia County and Otero County.

Paravilla flavipilosa (Cole)

Fig. 18

Villa flavipilosa Cole, 1923: 303. (Type in California Academy of Sciences.). *Villa (Paravilla) flavipilosa* (Cole), Painter and Painter, 1965: 439. *Paravilla flavipilosa* (Cole), Hull, 1973: 392.

Very much like *P. mercedis* and *P. xanthina*. *P. flavipilosa* is separated from *P. xanthina* by the much paler-yellow pile and tomentum, and from *P. mercedis* by the pale-yellow (not white) vestiture. The wings, except for the yellowish costal cell and basal half of first basal cell, are hyaline.

Male. - Black; basal antennal segments, lower half of front, face, palpi, legs except coxae and tips of tarsi, sides of abdominal dorsum, posterior margins of sternites fulvous.

Front with mixed black and pale-yellow hair, pale-yellowish to whitish tomentum. First antennal segment broadened on inner apical margin, twice length of second segment, with pale-yellow hair; second segment twice as wide as long, with short black hair; third segment long conical at base, one and one-half times longer than two basal segments combined, apical styliform portion about half length of conical basal portion. Face with short black hair above, yellow below at sides, whitish to pale-yellow tomentose. Palpus with yellow hair. Side of occiput whitish tomentose, becoming more yellow at vertex; cervical fringe pale yellow.

Thorax, scutellum, and abdomen pale-yellowish pilose and tomentose, abdominal segments two, three, and four with small median triangles of black scales on the posterior margins. Legs pale-yellowish pilose and tomentose, femora with scattered, small, black setulae; bristles black.

Genitalia with epandrium almost as wide as long, lower apical corner rounded, lower margin straight or nearly so; epiphallus curved, tip recurved, apex rounded, lower margin straight; in dorsal view, epiphallus with apex rounded, sides subparallel, slightly convex medially; aedeagus half length of epiphallus, pointed downward; basistylus short, broad, apex rounded; dististylus long, hooked apically.

Female. - Like male, but without black hair on front and face.

Variation. - Some males lack black hair on the head and some females have black hair on the front and face. Black tomentose triangles on the abdominal dorsum are not present in all specimens. In a few specimens the fore femora are lightly dusky.

Distribution. - This species was described from Baja California del Norte. Additional specimens have been studied from southern California, southern Arizona, and southwestern Texas. The species range extends north to Boulder, Colorado. The study of additional specimens has extended the range of distribution from Boulder, Colorado, south to Nuevo León, Mexico.

I have collected several specimens while they hovered over nests of *Diadasia rinconis rinconis* Ckll.; others have been reared by Dr. Ellen Ordway from the cells of that bee.

Paravilla floridensis Hall, new species

Fig. 19

This is possibly the most distinctive member of the genus, due to its black pile and tomentum and the snow-white tomentose markings on the abdominal dorsum. In addition, the post-alar and scutellar bristles are black, as are the face and antennae. Among the congeners, *P. floridensis* is possibly most closely related to *P. separata*.

Male. - Black; apical half or less of femora, all of tibiae ferruginous, oral margin narrowly so.

Front with black hair and dark-brown to black scales (at lower magnification, the scales reflect pale yellow), scales not reaching ocellar tubercle. Antennae with black hair on first two segments, first segment one and one-half times longer than second segment, widened apically; second segment a little wider than long; third segment with base short conical, apical styliform portion twice length of basal portion. Face with black hair and scales (scales reflect pale yellow in bright light). Proboscis short; at most, labella projecting beyond oral margin. Palpus with black hair. Side of occiput with whitish to pale-yellow scales, black hair only at vertex, scales absent; cervical fringe black.

Mesonotum with black hair, brownish hair on anterior margin, white hair above base of wing and at side between wing base and humeral callus; tomentum black, brownish anteriorly, tending toward ferruginous along side, scattered, black, bristlelike hair between wing base and humeral callus; pre-alar bristles fulvous, post-alar bristles black. Propleuron with fulvous and black hair; mesopleuron white-haired, with a few black hairs present on lower margin; metapleuron with white hair; sternopleuron with black hair and white tomentum, latter on upper portion only. Coxae and femora with black hair, scales on femora and tibiae pale, translucent.

Halter stem brown, knob creamy white. Squama brown,

with fringe of testaceous scales. Scutellum black pilose and tomentose; bristles black.

Basicosta with black scales and setulae. Wing with basal half black, apical half hyaline; color runs diagonally basad in an irregular line from tip of subcostal vein to hind margin of wing in basal fourth of axillary cell, both basal cells infuscated entirely, color in discal cell not reaching r-m crossvein except along vein M_1, color in anal cell reaching beyond base of fourth posterior cell, basal three-fourths of axillary cell colored; crossvein at base of second posterior cell convoluted.

Abdominal dorsum black pilose and tomentose, white hair present at sides on base of segments one and two; white tomentose spots at sides near bases of segments three to six, those on segments five and six expanded inward, but not reaching middle of segments; lateral margin of abdomen black pilose, white hair on segment one and on basal half of segment two. Venter of abdomen with white hair and scales on segments one and two and at side of third, rest of venter with ferruginous to brown scales and black hair.

Genitalia with epandrium narrow, length twice median height, lower apical corner bluntly projecting, lower margin concave; epiphallus humped, tip curved upward, apex wrinkled, rounded, lower margin convex; in dorsal view, epiphallus with sides narrowed below apex, then broadened basally, tip rounded; aedeagus straight, extending beyond mid-length of epiphallus; basistylus long, rounded apically; dististylus long, hooked apically.

Female. - Except for denser tomentum, female does not differ from male.

Holotype male and allotype. - From Archibold Biological Station, Highlands County, Florida, X.9.64 (P. H. Arnaud, Jr.). Both in collection of California Academy of Sciences.

Paratypes. - UNITED STATES: Florida: 13, topotypic, collected with types; 6, topotypic, X.21.65 (R. W. Dawson); Alachua County, 1, IV.12.55 (R. A. Morse); Levy County, 2, IV.13.55 (R. A. Morse), at flowers of *Erigeron quercifolius*.

Paravilla fulvicoma (Coquillett)

Fig. 20

Anthrax edititia var. *fulvicoma* Coquillett, 1887: 176. (Type in U.S. National Museum.). *Villa (Paravilla) fulvicoma* (Coq.), Painter and Painter, 1965: 439. *Paravilla fulvicoma* (Coq.), Hull, 1973: 392.

P. fulvicoma poses several problems in identification, among which are the fact that its appearance is almost identical to that of *P. edititoides*, and the large amount of variation that is encountered among the various populations. On the basis of a few individuals one would be tempted to subspeciate the variant populations, but when specimens are studied throughout the geographic range of the species it becomes apparent that there is no reliable means of distinguishing and/or defining these populations.

At the present time, the male genitalia offer the only reliable means of differentiating *P. fulvicoma* and *P. edititoides*. In *P. fulvicoma*, the aedeagus is narrow, straight, and nearly as long as the epiphallus; in *P. edititoides*, the aedeagus is broad, curved downward, and much shorter than the epiphallus. Supplemental differences may be found in the distribution and the time of occurrence: *P. fulvicoma* is found in more southwesterly regions and occurs in the spring and early summer; *P. edititoides* is more often found in the late summer and fall and occurs throughout the midwest and southwest. In general, *P. fulvicoma* has more black hair on the abdomen than does *P. edititoides*, and the basal third or more of at least the fore femur is usually dusky, as is the face; in *P. edititoides* these parts are entirely fulvous.

The following description is based in part on Coquillett's type series.

Male. - Black; first antennal segment, face, at least sides, legs except tarsi (basal third or more of fore femora often dusky), posterior margins of apical abdominal segments fulvous.

Front and face with black hair and dull dark-fulvous scales. First antennal segment twice length of second segment, not expanded appreciably at apex; second segment almost twice as wide as long, both segments with black hair; third segment a little longer than two basal segments combined, base long conical, styliform portion shorter than or equal to basal portion. Proboscis not projecting. Palpus with black hair.

Thorax and abdomen yellow to white pilose and tomentose, pile usually lighter above root of wing, disk of mesonotum with or without black hair; posterior margins of dorsal abdominal segments with or without one or two rows of black hairs, scutellum sometimes with a few black hairs in middle;

bristles fulvous to yellow. Coxae yellow to white pilose, sometimes also with a few black hairs; femora usually with black hair on posteroventral surface, a few specimens with hair and scales on the femora yellow.

Halter stem flavotestaceous, knob yellowish white.

Basicosta with fulvous scales and black setulae. Wing infuscated with reddish to yellowish brown on basal half; color runs diagonally basad from tip of subcosta to middle of axillary cell, color filling discal cell to or almost to r-m crossvein, both basal cells entirely colored; crossvein at base of second posterior cell bent slightly.

Genitalia with epandrium elongate dorsally, apex extending shieldlike over anus, lower apical corner rounded, lower margin concave; epiphallus long, narrow, straight, tip curved upward, apex rounded, lower margin straight; in dorsal view, epiphallus broad, rounded at tip, sides bowed outward; aedeagus long and narrow, more than half length of epiphallus, straight; basistylus long, narrow, apex rounded; dististylus short, narrow, hooked apically.

Female. - Essentially as described for male; usually, more black hair is present on disk of mesonotum and abdomen. Front in some specimens entirely fulvous pilose or mixed black and fulvous.

Variation. - Almost every character mentioned in the description above is subject to some variation. For the most part, the males have less black hair on the thorax and abdomen than do the females. Specimens from desert areas are often white pilose and tomentose; those from more humid climates are darker-yellow to fulvous pilose and tomentose. Specimens found along the coast or in areas of coastal climatic influence are usually fulvous in ground-color of the abdomen and legs, while those from drier areas are generally more black in ground-color.

Distribution. - Specimens have been studied from Washington south to Arizona in the West and from Texas to Kansas in the Midwest. From Arizona and Texas, the species ranges southward into eastern Mexico as far as Tamaulipas State. *P. fulvicoma* and *P. edititoides* have been taken together only once, in October, at Granite Dells, Yavapai County, Arizona.

A series of 14 small (6-9 mm in length) specimens from southern Arizona have the wings colored lightly and the legs entirely fulvous. I can find no differences between the male genitalia of this series and those of other *P. fulvicoma*.

Paravilla fumida (Coquillett)

Fig. 21

Anthrax fumida Coquillett, 1887: 177. (Type in U.S. National Museum.). *Anthrax obscura* Coquillett, 1894: 89, preocc. Macquart 1846, new synonymy. (Type in U.S. National Museum.). *Villa (Paravilla) fumida* (Coq.), Painter and Painter, 1965: 439. *Villa (Paravilla) coquilletti* Painter, 1965: 438 (new name for *obscura* Coquillett), new synonymy. *Paravilla coquilletti* (Paint.), Hull, 1973: 392. *Paravilla fumida* (Coq.), Hull, 1973: 392.

A direct comparison of the types of *P. fumida* (male) and *P. obscura* (female) showed them to be the same species. Therefore, *P. coquilletti* becomes an unnecessary change of name for *P. obscura* (Coquillett).

Among the smallest of the *Paravilla* species, measuring 5-8 mm in length, *P. fumida* is one of the easist to recognize because of the distinctive infumate coloring in the wings, almost entirely whitish or pale-yellowish pilose body, and shining integument, especially the scutellum.

Male. - Black; mesonotum, scutellum, and sometimes abdomen shining, first two antennal segments, side of face, palpi, legs except coxae and tips of tarsi, extreme lateral margin of abdomen, posterior margins of abdominal segments five to seven, nearly all of abdominal sternites, genital capsule flavotestaceous.

Eyes separated at narrowest point by twice width of ocellar tubercle. Front with black hair, tomentum fulvous. First antennal segment slightly longer than second segment, expanded apically on inner margin, a few light fulvous hairs on outer lower side of first segment; second segment almost twice as wide as long, both basal segments with black hair; third segment long conical at base, apical styliform portion about equal to length of basal portion. Face with short black hair and pale-fulvous tomentum, a few short fulvous hairs present at side of face below antennae. Palpus with yellow hair. Proboscis not projecting. Side of occiput white tomentose, becoming fulvous tomentose toward vertex, hair at vertex fulvous; cervical fringe pale fulvous.

Mesonotum pale-yellowish pilose, lighter above root of wings, disk with short black hair, tomentum and bristles fulvous. Pleura pale-yellow pilose and tomentose. Coxae with pale-fulvous hair and scales; femora with yellow hair on posteroventral surfaces, scales yellow; tibiae with yellow scales, bristles black.

Halter stem pale, knob white. Squama whitish, with fringe of white scales. Scutellum yellow pilose, pale-yellow tomentose, scales across base whitish; bristles yellow.

Basicosta with fulvous scales and black setulae. Wing infumate, color darker along anterior margin, longitudinal veins bordered with faint diffused color reaching margin of wing except at apex of wing, spots of diffused color at base of second submarginal and all posterior cells, all cells except costal and first basal hyaline to subhyaline medially, anal and axillary cells nearly entirely infumate; vein at base of second posterior cell straight, at right angles to veins M_1 and M_2; stumps of veins often present at knee of vein R_4.

Abdominal dorsum pale yellow to dark fulvous, lighter scales usually present across bases of second and third segments. Venter of abdomen light-fulvous pilose and tomentose.

Genitalia with epandrium short, higher than long, lower apical corner projecting slightly, lower margin straight; epiphallus straight, tip recurved, apex rounded, lower margin straight; in dorsal view, epiphallus narrow, tip rounded, sides concave before middle, wider basally; aedeagus half length of epiphallus, pointed downward; basistylus narrow, apex rounded; dististylus small, narrow, hooked apically.

Female. - Except for more abundant tomentum, does not differ from male.

Variation. - The pile and tomentum varies from light-yellow to nearly white, to dark fulvous. A few specimens may have some black hair on the apical abdominal segments. Wing coloring varies from quite dark to light.

Distribution. - This species is apparently confined to California. Ranging from the southern deserts north to Tuolumne County, it is on the wing in spring and early summer. The record from Oregon listed in the Catalog of Diptera of North America is based on a single male of a similar, but undescribed, species.

Paravilla fumosa Hall, new species

Fig. 22

A coastal species resembling *P. fumida*, but separated easily by the presence of black hair on the scutellum, femora, and posterior margins of most abdominal segments.

Male. - Small (7-10 mm in length). Black; integument of mesonotum and scutellum shining, lower half of front, first antennal segment, face above and below, legs except coxae, terminal abdominal segments fulvous to ferruginous.

Front with black hair, absent on lower fourth, scales dark testaceous above to fulvous near antennae. First antennal segment one and one-half times longer than second segment, widened on inner apical margin; second segment wider than long, both basal segments with black hair; third segment long conical basally, apical styliform portion only slightly longer than basal portion. Face with black hair and pale scales. Tips of labellae projecting beyond oral margin. Palpus brown, with yellow hair. Side of occiput pale-yellow to white tomentose, becoming fulvous toward vertex; cervical fringe white.

Mesonotum dark-yellow to fulvous pilose and tomentose, black hair on disk, side of mesonotum with whitish hair above root of wing, bristles fulvous. Pleura yellow pilose on anterior half, rest with white hair except for yellow metapleural tuft, tomentum white. Coxae with black hair and whitish scales; fore and mid femora with black hair, absent on hind femur, scales pale yellow; mid and hind tarsi with pale scales and black bristles.

Halter stem yellow, knob creamy white. Squama testaceous, with fringe of whitish scales. Scutellum with black hair in middle, fulvous hair at sides and across base, tomentum of light-fulvous scales, whitish scales across base; bristles fulvous.

Basicosta with fulvous scales and black setulae. Wing infuscated on basal half or more with dark brown, light spots of color at base of vein R_4 and at base of second posterior cell, darker areas of color at base of vein R_{2+3}, on r-m crossvein, and at bases of third and fourth posterior cells, cells other than costal and first basal within colored area with centers subhyaline, color in wing diffused, margin not delineated sharply; crossvein at base of second posterior cell somewhat sinuous.

Dorsal abdominal segments one to three white pilose,

rest with yellow hair, black hair in middle along posterior margins of segments two to four, that on five extending farther toward sides; tomentum white to whitish, semi-erect on second segment, appressed elsewhere, middle of posterior margins of segments two, three, and four with small inconspicuous spots of black tomentum; lateral margin of abdomen with white hair. Venter of abdomen white pilose, yellow hair on apical segments, a few black hairs present on sternites five and six, tomentum at base whitish, elsewhere testaceous to brown.

Genitalia with epandrium twice as long as median width, lower apical corner rounded, lower margin concave medially; epiphallus humped slightly, tip curved upward, apex rounded, lower margin convex medially; in dorsal view, epiphallus rounded apically, sides converging toward base; aedeagus more than half length of epiphallus, pointed downward; basistylus long, narrow, base broadly pointed, apex rounded; dististylus small, hooked apically.

Female. - Like male, except ground-color of abdomen entirely black, narrow bands of black scales across posterior margins of abdominal segments two to four, long black hair across posterior margins of all abdominal segments except the first. Basal half or less of the fore femur may be dusky.

Holotype male. - From Morro Bay, San Luis Obispo County, California, IX.14.56 (J. C. Hall); allotype from Pismo Beach, San Luis Obispo County, California, VIII.1.30 (T. F.

Winburn, R. H. Painter). Both in collection of U.S. National Museum.

Paratypes. - UNITED STATES: California: San Luis Obispo County, 1, topotypic, collected with type; 4, Baywood Park, IX.15.56 (J. C. and E. M. Hall); 1, Lake of Woods, VII. 19.31 (E. O. Essig); 2, Morro Dunes, IX.6.45 (A. L. Melander); 2, 4 mi. S Oceano, IX.12.56 (J. C. Hall).

I have three specimens (2 males, 1 female) from Wyoming in August, which differ slightly from the specimens found along the coast. Since they are so far from the type locality, I have elected not to include them in the type series.

Paravilla hulli Hall, new species

Fig. 23

P. hulli may be separated from the fall form of *P. syrtis* by the more extensive darkening on the face, denser tuft of black hair at the side of the second abdominal segment, and black ground-color of the venter of the abdomen. In the male, the lower apical corner of the epandrium is broad and pointed apically; in *P. syrtis*, the lower apical corner is rounded.

Male. - Black; first antennal segment, side of face, legs except coxae and tips of tarsi fulvous.

Front black pilose, flavotestaceous tomentose, tomentum not reaching ocellar tubercle. First two antennal segments with black hair; first segment one and one-half times longer than second segment, expanded on inner apical margin; second segment as long as wide; third segment almost twice length of segments one and two combined, base long conical, apical styliform portion about equal in length to conical basal portion. Face with black hair and fulvous scales. Proboscis not projecting beyond oral margin. Palpus with black hair. Side of occiput white tomentose, becoming yellowish tomentose toward vertex, small patch of fulvous scales behind ocellar tubercle; cervical fringe pale yellow.

Mesonotum with short black hair on disk, pale-yellow hair elsewhere, white hair above root of wing, tomentum pale yellow to white, bristles yellow. Pleura pale-yellow to white pilose, a few black hairs present on lower anterior portion of mesopleuron extending onto upper part of sternopleuron. Coxae with black hair and fulvous scales; fore and mid femora with black hair on posterior surface, absent on hind femur, tomentum on femora and tibiae fulvous, bristles and setulae black.

Halter stem fulvous, knob creamy white. Squama yellowish, with fringe of whitish scales. Scutellum with white hair at sides at base, rest with scattered black hair, tomentum whitish; bristles fulvous.

Basicosta with fulvous scales and black setulae. Wing infuscated on basal half, apical half hyaline; color runs diagonally basad in an irregular line from tip of subcostal vein to basal two-thirds of anal cell, thence into axillary cell along basal half of vein 2A, both basal cells colored, color in discal cell not reaching r-m crossvein except along vein M_1; crossvein at base of second posterior cell slightly sinuous.

Abdominal dorsum pale-yellowish pilose at base, white hair across basal half of second segment, short black hair across posterior margins of segments two to six, hair at apex of abdomen fulvous, bristlelike; tomentum whitish, black tomentose crossbands on segments two to four, reduced in extent posteriorly, becoming on four little more than a median triangular spot; side of abdomen with white hair at base, mixed black and yellow hair from apical corner of second segment to segment six, hair on seventh segment yellowish. Venter of abdomen with pale-yellow to whitish hair on basal three or four segments, rest with scattered black hair, tomentum flavotestaceous.

Genitalia with epandrium twice as long as high, lower apical corner pointed bluntly, lower margin slightly concave in middle; epiphallus humped, apex curved up, tip wrinkled, lower margin straight; in dorsal view, epiphallus narrowed beyond middle, apex rounded; aedeagus short, humped, half length of epiphallus; basistylus broad basally, apex almost truncate; dististylus short, hooked apically.

Female. - Does not differ appreciably from male.

Variation. - A few specimens have the abdominal tomentum patterned as in *P. syrtis*. In those specimens, the fulvous scales between the white and black scales are never as bright as in *P. syrtis*. Many females have the body, other than the black tomentose crossbands on the abdomen, entirely pale-yellow pilose and tomentose. Some females cannot be separated from the fall form of *P. syrtis*.

Holotype male and allotype. - From 2-10 mi. E Lamesa,

Dawson County, Texas, IX.13.65 (F. M. and M. C. Hull). Both in Canadian National Collection.

Paratypes. - UNITED STATES: Arizona: Cochise County, 2, 10 mi. W Wilcox, VIII.25.62 (F. M. Hull); Coconino County, 2, Flagstaff, X.2.57 (E. M. and R. H. Painter); 7, Grand Canyon, VIII.71 (F. M. Hull); Graham County, 1, Stockton Pass nr. Ft. Grant, VIII.31.47 (F. H. Parker). California: Inyo County, 2, Furnace Crk., Death Valley, IV.1.51. (J. W. MacSwain); San Bernardino County, 2, Phelan, V.9.49 (A. L. Melander); 4, upper Santa Ana River, IX.5.58, IX.15.56, IX. 15.46 (J. L. Sperry, A. L. Melander); 1, south fork Santa Ana River, VIII.2.68 (P. H. Arnaud, Jr.); 3, Verdemont, V.1. 46 (A. L. Melander); San Diego County, Borego, Palm Canyon, V.3.45 (A. L. Melander); Siskiyou County, 1, Gazelle, IX.22. 51 (J. C. Hall, E. I. Schlinger). Colorado: Logan County, 1, 16 mi. W Sterling, IX.2.64 (E. M. and R. H. Painter). Kansas: Morton County, 3, VI.02 (F. H. Snow). Mississippi: Lafayette County, 1, Oxford, VIII.7.43 (F. M. Hull). Nebraska: Kimball County, 4, Kimball, VI.26.61 (R. H. and E. M. Painter); Thomas County, 1, 2.5 mi. W Halsey, VII.16.67 (H. B. Leech). Nevada: Nye County, 14, Mercury, V.24.61. New Mexico: Guadalupe County, 1, 26 mi. W Santa Rosa, IX.7. 64 (F. M. and M. C. Hull); Lincoln County, 3, 2 mi. SW Capitan, IX.12.62 (R. H. and E. M. Painter); Otero County, 4, White Sands Natl. Monument, IX.10.65 (F. M. and M. C. Hull); Roosevelt County, 1, 14 mi. SW Elida, IX.13.62 (R. H. and E. M. Painter). Texas: Bexar County, 2, San Antonio, IX.17. 21 (R. H. Painter); Brewster County, 1, 40 mi. S Alpine,

IV.18.58 (R. H. and E. M. Painter); Brown County, 3, Brownwood, IX.11.20 (R. H. Painter); 2, Palo Duro Canyon, IX.30.57 (R. H. and E. M. Painter); Dawson County, 26, topotypic, collected with types; Dimmit County, 6, Catarina, IX.28.53 (R. H. and E. M. Painter); Glasscock County, 4, 9 mi. E Garden City, IX.8.58 (R. H. and E. M. Painter); Midland County, 1, Midland, VIII.24.50 (R. H. Painter); Mitchell County, 1, Colorado City, IX.14.21 (R. H. Painter); Runnels County, 6, Ballinger, X.18.57 (E. M. and R. H. Painter); 1, Winters, IX.15.21 (R. H. Painter); Taylor County, 1, Merkel, IX.15.21 (R. H. Painter); Uvalde County, 1, Uvalde, X.5.53 (A. W. Lindquist). Utah: Kane County, 1, Kanab, VIII.27.57. Wyoming: Carbon County, 1, Rock Springs, VIII.6.50 (R. R. Driesbach, R. H. Schwab); Teton County, 1, Zenith Entrance, Elk Refuge, VII.23.59 (F. M. Hull); Vinta County, 2, Cranston, VII.24.50. CANADA: Alberta: 1, Lethbridge, VIII.21.12 (J. B. Wallis). MEXICO: Baja California del Norte, 1, San Felipe, III.6.63 (P. H. Arnaud, Jr.); Jalisco, 5, 7 mi. E Ojuelos, VIII.10.63 (R. H. and E. M. Painter); Zacatecas, 1, 26 mi. NW Zacatecas, IX.3.62 (R. H. and E. M. Painter).

Paravilla imitans Hall, new species

Fig. 24

A moderate-sized (10-15 mm in length), dark species that resembles *P. macneilli* and *P. borea*. *P. imitans* can be separated from *P. borea* rather easily, by the black scutellar bristles. From *P. macneilli*, *P. imitans* can be differentiated primarily by the presence of tufts of black hairs on the sides of the second abdominal segment in both sexes, and by the more darkly colored wings.

Male. - Black; lower part of face and oral margin fulvous.

Front and face with black hair and pale-yellow scales. Antenna with black hair; first segment twice length of second segment, expanded on inner apical margin; second segment one and one-half times wider than long; third segment with base long conical, apical styliform portion longer than basal portion. Proboscis not projecting. Palpus with black hair. Side of occiput whitish tomentose, becoming yellow toward vertex, scattered, short, black hair on vertex; cervical fringe whitish.

Disk of mesonotum with thin, scattered, black hair, longer toward scutellum, anterior margin with pale-yellow pile, white hair above root of wing, tomentum pale yellow, not dense; pre-alar bristles fulvous, post-alar bristles black, or black basally and fulvous toward tips. Pleura pale-yellow to white pilose, scattered black hair on propleuron, lower margin of mesopleuron, and sternopleuron; whitish tomentum on upper part of sternopleuron. Legs with black hair and black scales.

Halter stem testaceous, knob whitish. Squama testaceous, with fringe of white scales. Scutellum with black hair, tomentum pale yellowish to whitish; bristles black, one or two fulvous bristles present on each side.

Basicosta with whitish and black scales and black setulae. Wing infuscated, apex and posterior margin hyaline, outline of color indistinct; color fills discal cell to r-m crossvein, faint spot of color at base of vein R_4, crossvein at base of second posterior cell infumate, color in anal cell reaching hind margin of wing, both basal cells colored; crossvein at base of second posterior cell straight or nearly so.

Abdominal dorsum white pilose, short black hair across posterior margins of segments two to six, hair at apex of abdomen fulvous; tomentum white, median triangles of black scales on segments two to five; lateral margin of abdomen white pilose on segments one through four, yellow hair on segment seven, tuft of black hair on apical corners of second, third, and fourth segments, fifth and sixth segments all black-haired. Venter of abdomen white pilose and tomentose, with scattered black hair from segment four to apex.

Genitalia with epandrium long, twice as long as wide, lower apical corner prolonged for a distance about equal to basal width of prolongation, lower margin concave slightly; epiphallus broad, slightly humped, tip curved upward, apex truncate, wrinkled, lower margin straight; in dorsal view, epiphallus broad, sides concave below middle, apex rounded; aedeagus short, less than half length of epiphallus, curved downward; basistylus narrow, about half as wide as long; dististylus long, hooked apically.

Female. - Except for the slightly more fulvous groundcolor of the face and more extensive black tomentose crossbands on the second and third abdominal segments, the female does not differ appreciably from the male.

Holotype male and allotype. - From Red Rock Canyon, Kern County, California, IX.26.57 (J. C. Hall). Both in collection of U.S. National Museum.

Paratypes. - UNITED STATES: Arizona: Pima County, 2, Silver Bell Bajada (J. L. Neff). California: Riverside County, 5, topotypic, collected with types; 1, 10 mi. W Blythe, X.2.54 (J. C. Hall); 1, Twentynine Palms, IX.5.46 (P. H. Timberlake).

Paravilla inatra Hall, new species

Fig. 25

This species belongs with the edititoides group of species; the male genitalic structure is similar to that of *P. fulvicoma*. In habitus, other than the infuscated wings, *P. inatra* resembles *P. flavipilosa* and *P. cinerea*. *P. inatra* may be recognized by the pale-yellowish to whitish tomentum on the entire body and the light yellowish-brown on the basal half of the wings.

Male. - Black; lower half of front, first two antennal segments, face, legs, and apical abdominal segments fulvous, posterior calli and middle or more of scutellum often dark reddish.

Front with fulvous hair and pale-yellow tomentum, scattered black hair present. First segment of antenna a little more than twice length of second segment, expanded on inner apical margin, with black and fulvous hair above, fulvous below; second segment twice as wide as long, with black hair; third segment longer than two basal segments combined, basal half long conical, apical styliform portion as long as basal portion. Face with short yellow hair and pale-yellow scales, short, scattered, black hair on side of face. Proboscis short; labellae projecting beyond apex of oral margin. Palpus with yellow hair. Side of occiput white tomentose, yellowish scales and hair at vertex; cervical fringe pale yellow.

Mesonotum pale-yellow pilose, hair on margins longer and paler than that on disk, black hair occasionally present on disk, hair along lateral margin above root of wing white to whitish, bristles yellow. Pleura whitish pilose and tomentose. Legs with white to pale-yellow hair and scales.

Halter stem testaceous, knob creamy white. Squama yellowish, with fringe of pale-yellow scales. Scutellum with somewhat appressed yellowish hair and white scales; bristles pale.

Basicosta with scales fulvous above, white below, setulae black. Basal half of wing tinged lightly with brownish-yellow; color extends from tip of vein R_1 diagonally basad to middle of axillary cell, margin jagged, not straight, color in discal cell reaching r-m crossvein along vein M_1 only, both basal cells colored, middle of second basal cell sometimes subhyaline, all but apical fourth of anal cell colored; vein at base of second posterior cell sinuous.

Dorsum of abdomen entirely pale-yellow to whitish pilose and tomentose. Venter of abdomen entirely whitish pilose and tomentose.

Genitalia with epandrium narrow, projecting shieldlike

dorsally over anus, lower apical corner rounded, lower margin sinuous; epiphallus straight, tip curved upward at about 15°, apex blunt, lower margin straight; in dorsal view, tip of epiphallus broadly rounded, sides subparallel; aedeagus narrow, long, straight, more than half length of epiphallus; basistylus broad at base, apex bluntly pointed; dististylus long, not appreciably hooked at apex.

Female. - Front entirely fulvous, basal conical portion of third antennal segment longer than apical styliform portion, black hair absent on face except at tip of oral margin, tomentum in general denser; otherwise as described for male.

Variation. - In some specimens, the reddish color of the post-alar calli and scutellum is bright and quite extensive; in others, not a trace of red is evident.

Holotype male. - From Tucson, Pima County, Arizona, V.13.61 (R. H. and E. M. Painter); allotype from 24 mi. N Tucson, Pima County, Arizona, VI.9.61 (R. H. and E. M. Painter). Both in collection of U.S. National Museum.

Paratypes. - UNITED STATES: Arizona: Cochise County, 1, V.20.61; Gila County, 1, Globe, V.10.36 (F. H. Parker); Pima County, 2, topotypic, collected with type; 20, collected with allotype; 1, Baboquiveri Mountains, IV.27.47 (A. L. Melander). California: Riverside County, 4, Deep Canyon, trail from Hwy. 74 to Horsethief Creek, V.30.73 (A. B. Tabet); San Bernardino County, 1, 8.5 mi. S Ivanpah, IX.13.55 (C. D. MacNeill). New Mexico: Otero County, 1, White Sands, VIII.50 (F. M. Hull). MEXICO: Sonora, 2, 23 mi. N. Ciudad

Obregón, IV.25.61 (R. H. and E. M. Painter); 6, 4 mi. S Magdalena, IV.19.61 (R. H. and E. M. Painter).

Paravilla lacunaris (Coquillett)

Fig. 26

Anthrax lacunaris Coquillett, 1892: 185. (Type in U.S. National Museum.). *Villa (Paravilla) lacunaris* (Coq.), Painter, 1933: 10; Painter and Painter, 1965: 439. *Paravilla lacunaris* (Coq.), Hull, 1973: 392.

Similar to *P. syrtis*. The two may be separated by the fulvous posterior half of the abdomen and the hyaline to subhyaline second basal cell in the wing in *P. lacunaris*, and in the diffenence in configuration of the male genitalia.

Male. - Moderate size, 10-12 mm in length. Ground-color fulvous; vertex and upper half of front, occiput, thorax, scutellum, coxae, base of abdomen, tarsi black.

Front with black hair and fulvous to white scales. Basal antennal segments with black hair; first segment of antenna almost three times length of second segment, not appreciably enlarged on inner apical margin; second segment one and one-half times wider than long; third segment long conical at base, rather evenly tapering to small styliform apex, segment about twice as long as two basal segments combined. Face with fulvous and black hair, tomentum light fulvous to yellowish. At most, labellae projecting beyond tip of oral margin. Palpus black, with black hair. Occiput whitish tomentose laterally, shading to yellowish tomentose and pilose at vertex; cervical fringe white.

Mesonotum pale-yellow pilose, lighter along lateral margin above root of wing, disk with short black hair, tomentum dense, pale yellowish, bristles pale. Pleura pale-yellow pilose, tomentum, where present, white. Coxae with black and fulvous hair, white to fulvous scales; fore and mid femora with short fulvous hair on posteroventral surface, hair absent on hind femur, scales fulvous, a few dark scales present on apical third of hind femur on anterior surface; scales on tibiae fulvous to light yellow, bristles black.

Halter stem flavotestaceous, knob whitish. Squama fulvous, with fringe of whitish scales. Scutellum yellow pilose and white tomentose; bristles pale.

Basicosta with fulvous scales and black setulae. Basal half of wing infuscated, apical half hyaline; color extends diagonally basad from tip of subcostal cell to apex of anal cell, margin of color not straight, color in discal cell reaching r-m crossvein, basal portion of wing lightly colored to midpoint of basal cells, nearly hyaline, outer margin of color darker, giving wing a banded appearance; crossvein at base of second posterior cell bent slightly.

Abdominal dorsum pale-yellowish pilose, tending toward white basally and fulvous apically, without black hair; tomentum whitish, black tomentose triangle on posterior margin of second segment not reaching lateral margin of abdomen; side of abdomen entirely pale fulvous or yellowish pilose. Venter of abdomen fulvous pilose and tomentose.

Genitalia with epandrium elongate, lower apical corner pointed, projecting slightly lower margin almost straight; epiphallus straight, curved upward at apex at 90° angle, lower margin almost straight, upper margin with short hair;

in dorsal view, epiphallus with apex rounded, sides narrowed medially, basal two-thirds or more with short hair; aedeagus short, half length of epiphallus, curved downward; basistylus elongate, rounded basally and apically, narrowed at apex; dististylus narrow, hooked apically.

Female. - Very similar to male, except for front with mostly fulvous hair, black hair across posterior margins of segments four to seven.

Variation. - Although the species is uniform in appearance, the following variations have been noted: abdomen may be entirely fulvous except for the first segement, or with only the three apical segments fulvous in ground-color; some females may have the venter of the abdomen with dusky rather than fulvous ground-color; the cells at the base of the wing may be clear hyaline or slightly yellowish; pile and tomentum vary from white to yellow fulvous.

Distribution. - At present, *P. lacunaris* is known only from the desert areas of Riverside and San Bernardino counties in southern California.

Paravilla leucothoa (Wiedemann), new combination

Fig. 27

Anthrax leucothoa Wiedemann, 1830: 638. (Type in Naturhistorisches Museum, Vienna.). *Anthrax diana* Williston, 1901: 280, preocc. Walker, 1849. *Villa (Paravilla) dianella* Painter, 1962: 101, n. n. for *diana* Will., new synonymy.

An examination of the types of *P. leucothoa*, *P. diana*, and *P. dianella* showed them to be the same species; therefore *P. dianella* was an unnecessary change of name.

This is a Mexican species belonging to the group that

has dark spots of color at the bases of the second submarginal and second posterior cells, and infuscated basal half of the wing. It may be distinguished from the congeners by the black femora, light-yellow body hairs, and white to silver scales on the sixth and seventh abdominal segments.

Painter and Painter (1962, p. 104) gave a key to five species that have spots of color at the apex of the wing. They separated *P. dianella* and *P. leucothoa* on the basis of the outer part of the wing being faintly smoky or hyaline and the oral margin light brown or black. In the specimens before me, collected at the same time and place and determined as *P. dianella*, all four characteristics were noted. The faintly smoky appearance of the hyaline portion of the wing is dependent upon the illumination used when viewing the specimens. Because of the confusion as to the identity of *P. leucothoa*, it is redescribed here.

Male. - Black; lower part of face, tibiae ferruginous.

Front black pilose, golden tomentose. Antennal segments one and two with black hair, first segment twice as long as second segment, not appreciably widened at apex; second segment wider than long; third segment almost twice length of segments one and two combined, base long conical, apical styliform portion equal in length to conical base. Face with black hair and scattered golden scales, tip of oral margin with a tuft of overhanging stiff black hairs. Proboscis not extending beyond tip of oral margin. Palpus with black hair. Occiput yellow tomentose, a few yellow and black hairs present on vertex; cervical fringe yellow.

Mesonotum with long black hair, anterior margin and side with light-yellow hair, posterior margin mixed black and yellow pilose; tomentum scattered, yellowish, tomentum usually not evident, bristles pale. Pleura yellow pilose and tomentose, tomentum not dense. Coxae with mixed black and yellow hair, yellow tomentum; femora with long black hair on posteroventral surfaces, scales flavotestaceous, black scales present on anterior surfaces of all femora; tibiae with flavotestaceous scales and black setulae.

Halter stem brown, knob creamy white. Squama testaceous, with fringe of yellow scales. Scutellum with pale-yellow hair and bristles, tomentum absent.

Basicosta with yellow scales and black setulae. Wing two-thirds or more colored, rest hyaline; color extends from tip of vein R_1 diagonally basad to include all of marginal cell except bulbous apex, first submarginal cell to base of vein R_4, across first posterior cell to large spot at base of second posterior cell, base of discal cell colored to r-m crossvein, thence basad to include bases of third and fourth posterior cells, basal three-fourths of anal cell, upper basal half of axillary cell, both basal cells colored, many specimens with a small, faint spot of color at or near apex of first posterior cell, margin of color definite; vein at base of second posterior cell bent slightly.

Abdominal dorsum pale-yellow pilose, a few black hairs across posterior margins of segments three to six, tomentum yellow, black tomentose spot in middle of base of second segment, crossbands of black tomentum on posterior margins of segments two to five, crossbands not reaching lateral margin of abdomen, segments six and seven with silver-white scales; lateral margin of abdomen pale-yellow-haired, with scattered black hairs on posterior corners of segments three to six. Venter of abdomen yellow pilose, black tomentose, a few long, appressed yellow hairs across posterior margins of second and third segments and laterally on third segment to apex.

Genitalia with epandrium elongate, almost twice as long as wide, lower apical corner acute, projecting slightly, lower margin straight; epiphallus narrow, widened apically, somewhat curved, apex blunt, with two submedian dorsal forward-projecting toothlike projections, lower margin slightly concave; in dorsal view, epiphallus narrow, apex bluntly rounded, sides concave near middle; aedeagus short, less than half length of epiphallus, pointed downward; basistylus large, basal half rounded, apex bluntly rounded; dististylus small, hooked strongly at apex.

Female. - Does not differ appreciably from male.

Variation. - The degree of intensity of the yellow-colored hairs and tomentum varies from light to dark; the black tomentose crossbands on the abdomen in some specimens are reduced to elongated triangles on the posterior margins of the segments; the coxae may be as described or entirely yellow pilose; apices of femora may be the same color as the tibiae.

Distribution. - Known only from Mexico, from Durango in

the north to Guerrero in the south. The flies are on the wing in August and September.

Paravilla lucida Hall, new species

Fig. 28

P. lucida is similar to *P. lacunaris*, differing from it by the absence of black tomentum on the second dorsal abdominal segment, and the almost entirely hyaline anal cell.

Male. - Black; lower half of front, first two antennal segments, face, legs except tips of tarsi, posterior margins of terminal abdominal segments fulvous.

Front black pilose, dull-fulvous tomentose, scales reaching ocellar tubercle. First antennal segment three times length of second segment, cylindrical, not expanded apically; second segment as wide as long, both basal segments with black hair; third segment one and one-half times as long as segments one and two combined, base long conical, tapering rather evenly to styliform apex. Face with black hair, scales fulvous. Proboscis not projecting. Palpus with black hair. Side of occiput with pale-yellow scales, darker at vertex, a few fulvous hairs at vertex; cervical fringe pale yellow.

Disk of mesonotum with short black hair, pale fulvous hair on anterior and posterior margins, stripe of whitish hair above root of wing; tomentum pale fulvous, hairlike, not dense, bristles fulvous. Pleura pale-yellow to white pilose and tomentose.

Coxae with pale-fulvous hair and whitish tomentum, a

few black hairs present toward apices of coxae. Fore femur with black hair, mid femur with yellow and/or black hair, hair absent on hind femur; scales light fulvous on all femora and tibiae, with darker but not black scales on anterior surfaces of all femora; tibiae with black setulae and bristles.

Halter stem flavotestaceous, knob yellowish white. Squama flavotestaceous, with fringe of pale-yellow scales. Scutellum with short black hair, a few white hairs present across base, tomentum fulvous, whitish across base and at sides; bristles fulvous.

Basicosta with pale-fulvous scales and black setulae. Wing hyaline, infuscated in costal, subcostal, basal half of marginal, all of first basal, basal half of discal, extreme bases of third and fourth posterior, extreme base and apex of second basal, and base of anal cells; light color extends along anal vein to a little beyond posterior crossvein, color in discal cell reaching r-m crossvein except in middle; vein at base of second posterior cell slightly sinuous.

Abdominal dorsum pale-yellow pilose, posterior margins of segments two to five with short black hairs; tomentum dense, pale yellow across bases of segments two and three, darker fulvous elsewhere, fulvous scales lighter on apical segments; lateral margin of abdomen entirely pale-yellow pilose. Venter of abdomen whitish pilose and tomentose, pile and tomentum becoming more yellowish apically.

Genitalia with epandrium short, about twice as long as median height, lower apical corner acute but not prolonged, lower margin concave slightly; epiphallus straight, broad, tip curved upward, apex truncate, wrinkled, lower margin convex beyond middle; in dorsal view, epiphallus short, apex truncate, sides concave below middle; aedeagus short, not reaching mid-length of epiphallus, pointed downward; basistylus long, narrow, base and apex bluntly pointed; dististylus small, hooked apically.

Female. - As described for male, except for denser tomentum and fewer dark scales on anterior surfaces of femora.

Holotype male and allotype. - From Rancho Santa Margarita, 20 mi. S El Arco, Baja California del Sur, Mexico, V.13. 61 (W. W. Gibson), at flower of *Cryptantha*. Both in collection of U.S. National Museum.

Paratypes. - MEXICO: Baja California del Sur, 17, topotypic, collected with types; 1, San Quintín, VII.7.25 (H. H. Keifer).

Paravilla macneilli Hall, new species

Fig. 29

This is one of the few species of *Paravilla* in which sexual dimorphism is conspicuous. The males have the abdomen almost entirely white pilose and tomentose, while the female has abundant black hairs on the abdomen, as well as broad black tomentose crossbands on segments two to four.

P. macneilli resembles *P. fumosa* in that both have infumate wings without a definite margin to the color. The two species are separated by the much darker legs and presence of black tomentose crossbands on the abdominal dorsum in

P. macneilli. The spot of color at the base of vein R_4 and at the base of the second posterior cell are not sharply defined in either species.

Male. - Black; first antennal segment, oral margin narrowly fulvous.

Front and face with black hair and fulvous scales, scales on front not reaching ocellar tubercle. Antennal segments one and two with black hair; first segment twice length of second, expanded on inner apical margin; second segment almost twice as wide as long; third segment one and one-half times length of two basal segments combined, base broad, long conical, apical styliform portion about one-third length of broad basal portion. Proboscis not projecting. Palpus with black hair. Side of occiput with light-yellow scales, vertex with short black hair, patch of black scales behind ocellar tubercle; cervical fringe pale yellow.

Mesonotum black pilose, yellow hair on anterior margin and sides, a few yellow hairs present across posterior margin, tomentum absent, bristles mixed black and yellow. Pleura pale-yellow pilose and tomentose. Hair on coxae mixed yellow and black, tomentum, where present, yellow; femora with black hair and pale-yellow scales; scales on tibiae slightly darker yellow than on femora.

Halter stem testaceous, knob whitish. Squama brown, with fringe of whitish scales. Scutellum black pilose, tomentum yellowish, scattered; bristles black.

Basicosta with black and light-yellow scales and black setulae. Wing infuscated on basal half or more, color not margined, middle of cells in infuscated portion subhyaline; color extends to posterior margin of wing in anal cell and along posterior margin in axillary cell, spots of color on base of vein R_4 and on vein M_2, color not dark but always present; knee of vein R_4 usually with a short stump of vein.

Abdominal dorsum white pilose and tomentose, black hair on posterior margins of segments four to seven; middle of posterior margins of segments three and four with small, triangular, black inconspicuous, tomentose spot; lateral margin of abdomen white pilose, with a few black hairs present on sides of segments five to seven. Venter of abdomen with pale-yellow hair on segments one to three, black hair on segments four to seven, tomentum pale yellow.

Genitalia with epandrium short, broad, apex rounded, lower margin concave; epiphallus short, thick, convex dorsally, tip curved upward, lower margin convex near middle; in dorsal view, tip of epiphallus rounded, sides narrowed basally; aedeagus half length of epiphallus, pointed sharply, bent down slightly; basistylus bluntly acute apically.

Female. - Scattered pale scales on mesonotum, white hair above and below base of wing, thoracic bristles yellow, scutellar bristles yellow; abdomen with broad bands of black hair across posterior margins of all segments except the first, tomentum pale yellowish with broad crossbands of black scales on posterior halves of segments two to five, that on segment four occupying almost the entire segment; lateral margin with black hair on second and following segments; tomentum on venter darker, more brownish. Otherwise as described for male.

Holotype male and allotype. - From near Mono Pass, Inyo County, California, 1200 ft. VIII.14.57 (C. D. MacNeill). Both in collection of California Academy of Sciences.

Paratypes. - UNITED STATES: Arizona: Pima County, 1, Elkhorn Ranch, Baboquivari Mtns., IV.21.62 (M. L. Muller). California: Inyo County, 11, topotypic, collected with types; Mono County, 4, Tioga Pass, Hall area, VIII.1.69 (A. Moldenke); Siskiyou County, 1, Siskiyou Natl. Forest, VII.14.35 (R. H. Beamer). Colorado: Delta County, 1, 8 mi. N Cedaridge, VII.3.59 (R. H. and E. M. Painter). Wyoming: Moran County, 1, Jackson Hole Biological Station, VII.16-30.61 (H. E. Evans).

In the Canadian National Collection is a single male specimen from Mexico that is tentatively placed with this species. It is quite atypical in that the wings are infuscated darkly, much as in *P. leucothoa*; otherwise it conforms to the other specimens of *P. macneilli*. It is not considered a paratype.

Paravilla mercedis (Coquillett)

Fig. 30

Anthrax mercedis Coquillett, 1887: 166. (Type in U.S. National Museum.). *Villa mercedis* (Coq.), Cole, 1923: 299. *Villa (Paravilla) mercedis* (Coq.), Painter and Painter, 1965: 439. *Paravilla mercedis* (Coq.), Hull, 1973: 392.

This species resembles *P. flavipilosa*. It differs by the hair and scales being white or whitish and by the presence of black scales on at least the anterodorsal surfaces of the femora. In addition, the epiphallus of the male is thick and straight, while the male of *P. flavipilosa* has the epiphallus narrow and humped. There is also some resemblance to *P. cinerea*, from which *P. mercedis* is distinguished by the hyaline wings.

Male. - Black; most of front, two basal antennal segments, face, palpi, lower half of occiput, legs including coxae, posterior margins of abdominal segments, nearly all of apical abdominal segments fulvous to rufous. In some specimens, the scutellum is vaguely reddish.

Front yellow pilose and tomentose, with black hair at ocellar tubercle and along eye margins on upper half of front. First antennal segment twice as long as second, expanded apically, with yellow hair; second segment twice as wide as long, with short black and longer yellow hair; third segment longer than two basal segments combined, base long conical, about equal in length to apical styliform portion. Face with yellowish hair and whitish-yellow scales. Palpus with yellow hair. Occiput white tomentose.

Mesonotum pale-yellowish pilose, white tomentose, tomentum dense, no black hairs on disk, bristles yellow. Pleural area white pilose and tomentose. Coxae with white to pale-yellowish hair and white scales; femora with yellow hair and scales, anterodorsal surface of all femora with black or dark-brown scales; tibiae with scattered yellow scales and black setulae, bristles black.

Halter stem and knob yellow. Squama pale yellow, with fringe of pale scales. Scutellum pale-yellow pilose, white tomentose; bristles yellow.

Scales at base of basicosta white, flavotestaceous elsewhere, setulae black. Wing hyaline, costal cell yellowish, veins at base of wing and along anterior margin yellowish, rest brown; vein at base of second posterior cell somewhat sinuous.

Abdomen entirely white or whitish pilose and tomentose.

Genitalia with epandrium in lateral view long and narrow, lower apical corner prolonged slightly, lower marign concave; epiphallus straight, thick, tip recurved slightly, apex rounded, wrinkled, lower margin convex slightly; in dorsal view, epiphallus with apex truncate, sides subparallel, diverging slightly basad; aedeagus short, half length of epiphallus, pointed downward; basistylus long, somewhat pointed on both ends; dististylus short, not appreciably hooked at apex.

Female. - Does not differ appreciably from male.

Variation. - The color of the hair and sometimes the scales vary from white to yellow. A few female specimens have been seen that have very short black hair on the disk of the mesonotum. Many specimens have the legs brownish-red rather than fulvous, and a few have the femora nearly entirely covered with black or dark-brown scales.

Holotype. - From Merced County, California.

Distribution. - I have seen specimens from California, Arizona, and New Mexico. The record from Colorado, given in the Catalog of Diptera of North America, is questionable. *P. mercedis* is, for the most part, a species inhabiting deserts or arid areas. It has been taken at flowers of *Petalonyx thurberi* and *Lepidospartum* species. It is on the wing from April to August; the majority of the specimens were taken in June and July.

Paravilla mexicana Hall, new species

Fig. 31

A rather large (up to 15 mm in length) species with somewhat infumate wings. The abdomen is banded with yellow and black and the apex is silver white; the antennae and legs are black. In general appearance, *P. mexicana* is much like *P. leucothoa*. The two may be separated mainly by the more infumate wings in *P. mexicana*; also, the color in the wing is not margined and there is no isolated spot of color at the base of vein R_4.

Male. - Black; lower part of face and oral margin ferruginous or testaceous.

Front and face with black hair and shining yellow scales. First antennal segment cylindrical, not widened apically, twice length of second segment, which is as long as wide, both basal segments with black hair; third segment almost a third longer than two basal segments combined, base long conical, apical styliform portion longer than basal portion. Tips of labellae projecting beyond oral margin. Palpus with black hair. Occiput with yellow hair and scales; cervical fringe yellow.

Disk of mesonotum with long black hair. longer toward scutellum; rest of mesonotum with light-yellow hair, paler above root of wing; tomentum yellow, hairlike, not appressed, bristles yellow. Pleura pale-yellow pilose, a few dark hairs present on lower part of propleuron. Coxae and femora with black hair; femora and tibiae with black scales.

Halter stem testaceous, knob creamy white. Squama brown, with fringe of fulvous scales. Scutellum with black hair in middle, yellow hair at sides; bristles yellow.

Basicosta with black scales and setulae, a few yellowish scales present on lower margin at base. Wing infumate, color not sharply margined nor dark, costal cell yellowish brown, middle of marginal cell and basal part of first submarginal cell to base of vein R_4 and basal half of first posterior cell brownish, first basal cell tinged lightly with yellowish brown, second basal cell hyaline, diffused spots of color at base of vein R_4, base of second, third, and fourth posterior cells, discal cell with a faint coloring running from r-m crossvein to base of fourth posterior cell, anal and axillary cells hyaline except for diffused color along vein 2A; crossvein at base of second posterior cell sinuous.

Abdominal dorsum pale-yellow pilose, with black hair across median portions of posterior margins of segments two to six; tomentum on bases of segments two to five pale yellow, remainder of segments two to five with black tomentum, segments six and seven entirely silver-white tomentose; lateral margin of abdomen entirely pale-yellowish pilose. Venter of abdomen with black hair, pale-yellow hair on sternites one and two, tomentum brassy orange.

Genitalia with epandrium large, more than twice as long

as median width, lower apical corner extruded, bluntly rounded; epiphallus straight, tip curved upward slightly, apex rounded, lower margin convex, sides with winglike extensions; in dorsal view, epiphallus ovoid, sides concave toward base; aedeagus short, straight, pointed downward; basistylus large, rounded at base, narrowed apically, apex rounded; dististylus large, hooked apically.

Female. - As described for male except for denser tomentum.

Holotype male. - From 3 mi. N León, Guanajuato, Mexico, IX.17.59 (R. H. and E. M. Painter); allotype from 4 mi. N León, Guanajuato, Mexico, IX.23.65 (R. H. and E. M. Painter). Both in collection of U.S. National Museum.

Paratypes. - All, except as noted, collected by R. H. and E. M. Painter. MEXICO: Guanajuato: 4, Comanjilla, IX.23.63; 1, 27 mi. S Salamanca, IX.22.63. Guerrero: 2, 2.5 mi. S Iguala, IX.13.63; 1, 3 mi. N Mexcala, IX.13.63. Jalisco: 1, Puerta Barranquitas, 18 mi. NW Magdalena, X.10.75 (J. A. Powell, J. A. Chemsak). Michoacán: 2, 4 mi. W Morelia, IX.22.65. Morelos: 2, 16 mi. E Cuernavaca, IX.17.63. Oaxaca: 1, 43 mi. W Tehuantepec, VIII.22.63 (H. V. Weems, Jr.).

Paravilla mexicana was taken flying with an undescribed species of *Poecilanthrax* that is similar to *Poecilanthrax ceyx* (Loew). The two are similar in appearance and, in fact, were mixed in the collection as *Paravilla*.

Paravilla montivaga Hall, new species

Fig. 32

This species has in the past been confused with both *P. edititoides* and *P. emulata*. It is distinguished from *P. edititoides* by the faint spot of color at the base of the second submarginal cell. The separation from *P. emulata* is not obvious, other than in the male genitalia: in general, *P. montivaga* has lighter-brown coloring in the wing, with faint spots of color at the bases of the second submarginal and second posterior cells; in *P. emulata* these spots are quite pronounced. A few black hairs at the sides of the fourth and fifth abdominal segments are usually present in *P. emulata*, but are rarely present at these places in *P. montivaga*.

Male. - Black; basal antennal segments, face, legs except coxae fulvous.

Front and face with black hair and dull-fulvous scales. First antennal segment one and one-half times longer than second segment, widened apically, second segment a little wider than long, both basal segments with black hair; third segment twice length of two basal segments combined, base short conical, apical styliform portion two and one-half times longer than conical base. Proboscis short, not projecting beyond oral marign. Palpus brownish, with black hair. Occiput with pale-yellow scales and more fulvous, short, appressed hair; cervical fringe pale yellow.

Mesonotum with light-fulvous hair and tomentum, disk with short, scattered, black hair, light whitish hair above root of wing, tomentum hairlike, erect; bristles yellow. Pleura very pale-fulvous pilose, tomentum whitish. Coxae with mixed black and yellowish hair and yellowish tomentum; femora with black hair and numerous black setulae, scales on femora and tibiae light fulvous, bristles black.

Halter stem testaceous, knob flavotestaceous. Squama testaceous, with fringe of bright-orange scales. Scutellum light-fulvous pilose and tomentose, bristles yellow.

Basicosta with fulvous scales and black setulae. Wing with basal half or more infuscated, color not sharply margined, color in discal cell to r-m crossvein, almost all of anal cell colored, basal half of axillary cell colored, light or faint spots of color at base of vein R_4 and at base of second posterior cell (best seen in indirect light); vein at base of second posterior cell straight.

Abdominal dorsum and venter yellowish pilose and tomentose, tomentum on tergites three and five slightly lighter in color.

Genitalia with epandrium, except for petiole, almost square, lower apical corner bowed outward slightly, lower margin straight; epiphallus humped, tip curved upward, apex rounded, wrinkled, lower margin concave medially; in dorsal view, epiphallus with apex rounded shallowly, sides subparallel, converging below middle; aedeagus short, less than half length of epiphallus, curved downward; basistylus broad on basal half, apex bluntly rounded; dististylus long, narrow, hooked slightly at apex.

Female. - As described for male.

Variation. - Many specimens have yellow hair on front near ocellar tubercle and next to margins of eyes. The pile and tomentum vary from a very pale yellow to a dark, nearly fulvous yellow. In some specimens, the slightly paler scales on the abdomen are quite pronounced. The wings may be colored darkly or lightly, the spot of color at base of second posterior cell may be absent and the second basal cell may be subhyaline.

Holotype male and allotype. - From Old Station, Shasta County, California, VI.24.55 (J. R. Jessen, D. L. Dahlsten). Both in collection of U.S. National Museum.

Paratypes. - UNITED STATES: Arizona: Coconino County, 1, 10.8 mi. N Cedar Ridge, VIII.30.67 (J. C. Hall); Maricopa County, 2, 5 mi. E Mesa, IX.17.64 (F. M. and M. C. Hull); Navajo County, 4, Indian Pine, VIII.28.64 (M. E. Irwin); 1, S Arizona, VIII.02 (F. H. Snow). California: Alpine County, 2, Hope Valley, VII.18.48 (C. D. MacNeill); Butte County, 1, Feather River, VI.14.60 (M. Cazier, T. Aitken); El Dorado County, 1, Echo Portals, VIII.4.42 (P. H. Arnaud, Jr.); 1, Pyramid Lake Ranger Station, VIII.12.48 (C. Chan); Fresno County, 2, Huntington Lake, VII.28.19 (E. P. van Duzee); 1, Wood's Creek, VII.22.10 (E. C. van Dyke); Inyo County, 1, Argus Mtns., VI.22.37 (N. W. Frazier); Lassen County, 3, Summit Camp, VI.28.49 (H. A. Hunt); Mariposa County, 1, Yosemite Valley, VII.8.21 (E. C. van Dyke); Monterey County, 1, Bradley, IX.16.43 (F. M. Hull); Napa County, 1, Samuel Springs, V.20.53 (J. C. Hall); Nevada County, 2, Boca, VI.11. 59 (R. M. Bohart); 3, Sagehen, nr. Hobart Mills, VII.8.64

(R. E. Scott); 1, 4 mi. NW Hobart Mills, VII.5.51 (E. I. Schlinger); 4, Sagehen Creek, VI.25.66 (E. E. Grissell, C. J. Horning); Orange County, 3, Santa Ana, IX.17.64 (F. M. Hull); Placer County, 2, Angora Lake, VII.11.15; 1, Fallen Leaf Lake, Tahoe, VII.26.15 (E. C. van Dyke); 1, Fallen Leaf, VII.16.17 (J. M. Aldrich); 1, Globins, Lake Tahoe, VII.17.55 (E. A. Kurtz); Plumas County, 1, Buck's Lake, VII.1.49 (E. I. Schlinger); 1, Meadow Valley, VI.17.24 (E. C. van Dyke); 6, 4 mi. N Quincy, VII.2.49 (H. A. Hunt); 1, Portola, VII.2.54 (G. Schaefers); 1, Silver Lake, VI.19.24 (E. C. van Dyke); Riverside County, 1, Riverside, X.13.27 (P. H. Timberlake); San Mateo County, 4, Redwood City, VI.20.50, IX.15.57 (P. H. Arnaud, Jr.); Santa Clara County, 1, Palo Alto, IX.4.21 (E. P. van Duzee); Shasta County, 16, topotypic, collected with type; 2, 9 mi. W Hat Creek P.O., VII.3.55 (J. C. Wells); 1, 2 mi. SE Hat Lake, VIII.1.75 (T. Plichta); 3, Lake Eller, VII.9.47 (R. L. Usinger, D. W. Adams); 6, Mt. Shasta, VI.29.35 (E. I. Beamer); 1, Old Station, VI.21.55 (W. W. Middlekauff); 5, Shasta Springs, VI.8.20 (C. L. Fox); 7, Shasta County, VIII.26.21 (J. A. Kusche); Sierra County, 1, Allegheny, VII.11.58 (R. E. Darby); 5, Gold Lake, VII.29.21, VII.8.54 (C. L. Fox, R. H. Goodwin); 1, Sierraville, VI.28.66 (E. E. Grissell); Sonoma County, 1, Sobre Vista, VI.9.10 (J. A. Kusche); Trinity County, 1, Coffee Creek Ranger Station, VII.14.55 (A. J. Mueller); 2, Mountain Meadow Ranch, head Coffee Creek, VII.8-10.69 (W. G. Goodwin); Tulare County, 3, Long Meadow, Giant Forest, VII.21.29 (C. L. Fox); 1, Mineral King, VII.15.58 (W. E. Simmonds); Tuolumne

County, 3, Bumble Bee, VII.3.51 (J. C. Hall, J. C. Downey); 1, Cow Creek, VI.24.51 (E. I. Schlinger); 6, Dardanelles, VII.11.57 (J. W. MacSwain, A. J. Pritchard); 3, nr. Dardanelles, VI.26.51, VII.11.70 (C. A. Downing, T. Plichta); 3, Eleanor Lake, VII.2.57 (S. M. Kappos, C. A. Downing); 3, Pinecrest, VII.1.47 (P. H. Arnaud, Jr.); 10, Strawberry, VI.21.51, VII.1.52, VI.11.59 (J. C. Hall, J. W. MacSwain, W. H. Lange). Colorado: Delta County, 1, Paonia, VI.14.26 (E. C. van Dyke); Montrose County, 1, 7 mi. W Sapinero, VII.19.59 (R. H. and E. M. Painter). Idaho: Kootenai County, 1, Coeur de Alene, VII.16.63 (N. Jorgensen); 1, Bear Lake, VII.9.20; Nez Perce County, 1, top Lewiston Grade, VI.27.72 (W. B. Garnett); Oneida County, 1, Slayer Cow Camp, VII.14.71 (G. E. Knowlton); Shoshone County, 2, 1 mi. S Avery, VII.16.72 (R. F. Lagier). Oregon: Baker County, 17, Sparta, VII.2.22 (E. C. van Dyke); Harney County, 1, Steen Mtns., VI.25.22 (E. C. van Dyke); Klamath County, 4, Eagle Ridge, VI.4.24. (C. L. Fox); 3, Ft. Klamath, VII.17.71 (E. Villegas); 2, Soda Mtn., VII.7.59 (C. Fitch); 2, Sun Pass, VII.18.71 (E. Villegas); Umatilla County, 1, Freewater, IX.9.14 (G. F. Mozenette); Wallowa County, 1, Hurricane Creek, Wallowa Mtns., VIII.3.50. Utah: Cache County, 1, Blacksmith Fork, VII.1.64 (W. J. Hanson); Duchesne County, 1, Monarch, IX.3.57; 1, Wasatch Mtns., VII.8; Sanpete County, 1, Spring Hollow, VII.24.36 (R. E. Nye); Utah County, 1, State Canyon, Provo, VII.4.20 (T. Spalding). Washington: Austin County, 1, Field Springs State Park, VI.26.72 (A. N. Gartrell); Kittatas County, 8, Cliffdell, VII.7.35 (E. I. Beamer); Spokane

County, 1, McCroskey State Park, nr. Farmington, VII.16.70 (J. D. Akre); 1, Medical Lake, VII.14.20 (R. C. Shannon); Whitman County, 4, Smoot Hill, Pullman, VII.10.70 (H. S. Telford, J. Novak). Wyoming: Fremont County, 1, nr. Lander, VII. (R. Moodie); 1, 10 mi. NE South Pass, VIII.4.55 (G. E. Bohart). CANADA: British Columbia: 1, Single Creek Road, Keremeos, VI.15.33 (A. N. Gartrell). MEXICO: Chihuahua: 1, Cuiteco, IV.14.69 (T. A. Sears, R. C. Gardner, C. S. Glaser); 1, Santa Clara Canyon, 8 mi. W Parrita, IX.1.56 (D. D. Linsdale).

Paravilla nigriventris Hall, new species

Fig. 33

This species belongs to the consul-like group of species, in which the lower half of the body and legs are entirely black pilose and tomentose. In appearance, *P. nigriventris* is almost identical to *P. consul*; the two may be separated by the more extensive coloring in the axillary cell in the wings, the more pointed lower apical corner of the epandrium, and the non-curved apex of the epiphallus in *P. nigriventris*.

Male. - Black; stems of halteres creamy-white to yellowish.

Front, antennae, and face with black hair, front and face with scattered, dull golden-yellow scales. First antennal segment cylindrical, not expanded apically, twice length of second segment; second segment rounded, as wide as long; third segment nearly one and one-half times longer than first two segments combined, base short conical, styliform portion one and one-half times longer than basal portion. Palpus with black hair. Occiput with yellow scales, yellow hair at vertex; cervical fringe yellow on upper half, black on lower half.

Mesonotum with black hair on disk, longer toward scutellum, collar and lateral margins with pale-yellow to whitish hair; tomentum yellow, hairlike, dense, bristles pale. Pleura black pilose except for upper third, which has pale-yellow hair, tomentum on sternopleuron brownish. Legs with black hair and scales.

Squama white, with fringe of white scales. Scutellum with black hair and bristles, tomentum yellow.

Basicosta with black scales and setulae. Wing infuscated on basal half, apical half or more hyaline; color runs from near apex of subcosta diagonally basad in an irregular, but not stepped, line through discal cell opposite r-m crossvein to apical third of anal cell, thence into axillary cell and curving basad to base of axillary cell, color not reaching posterior margin of wing, both basal cells infuscated.

Abdominal dorsum white pilose on sides of segments one and two, rest with black hair; pale-yellow tomentose crossbands on bases of segments two through four, bands narrowed medially, not divided, segment five entirely black tomentose, segments six and seven white to silver-white tomentose, black tomentum elsewhere; lateral margin of abdomen with whitish hair on first segment and basal half of second segment, rest of margin black-haired. Venter of abdomen entirely black pilose and tomentose.

Genitalia with epandrium narrow, lower margin concave, lower apical corner projecting bluntly, the projection short, about as long as wide; epiphallus curved slightly, broad, tip not recurved, apex broad, blunt, wrinkled, lower margin straight or nearly so; in dorsal view, epiphallus broad, concave, apex rounded, sides rounded, converging at base; aedeagus half length of epiphallus, curved downward slightly; basistylus broad at base, apex rounded.

Female. - Except for usual sexual differences, does not differ from male.

Holotype male and allotype. - From 6 mi. N Iguala, Guerrero, Mexico, IX.15.63 (R. H. and E. M. Painter). Both in collection of U.S. National Museum.

Paratypes. - All, unless noted otherwise, collected by R. H. and E. M. Painter. MEXICO: Chiapas: 19, 7 mi. SW Centalapa, IX.2.63, IX.4.63, IX.6.63, IX.12.63; 1, 35 mi. E Comitán, IX.4.63; 18, 22 mi. SE Comitán, IX.4.63; 9, 23 mi. W Orozocantla, IX.6.63; 4, San Jacinto, IX.27.60 (C. H. Martin); 2, Tuxtla, Guitiarrez, IX.27.60 (C. H. Martin); 1, 13 mi. N Tuxtla, Guitiarrez, IX.6.63. Colima: 3, 5 mi. E Colima, VIII.27.62 (C. H. Martin). Guerrero: 2, Azizintla, IX.8.60 (C. H. Martin); 2, 2 mi. N Azizintla, IX.18.63; 11, 24 mi. N Chilpancingo, VIII.31.59; 6, 32 mi. N Chilpancingo, IX.19.60, VIII.31.59 (D. W. Martin); 4, 3 mi. N Chilpancingo, VIII.21.59; 1, nr. Iguala, IX.9.60 (C. H. Martin); 17, 6 mi. N Iguala, IX.15.63; 10, 3 mi. N Mexcala, IX.13.63. Michoacán: 1, 13 mi. S Morelia, IX.22.63. Morelos: 3, 15 mi. E

Cuernavaca, IX.17.63; 1, 32 mi. S Cuernavaca, IX.12.63; 7, 17 mi. S Cuernavaca, IX.12.62, IX.16.63; 1, 20 mi. S Cuernavaca, IX.12.63; 8, 24 mi. N Cuernavaca, VIII.31.59; 1, 2 mi. N Huajintlán, IX.18.63; 7, Lobo Canyon, Cuernavaca, IX.22.60, VIII.14.62; 1, Tempilpa Nuevo, IX.14.66 (A. Ortega). Oaxaca: 1, 5 mi. S Ejutla, VIII.31.63; 2, Huajuapán, Hwy. 190, IX.30.60 (C. H. Martin); 18, 3 mi. N Huajuapán, IX.6.59, IX.8.59, IX.2.63; 14, 1 mi. W Huajuapán, IX.11.63; 2, Las Catarinos, IX.28.60 (D. W. Martin); 5, 15 mi. W Tequisitlán, IX.9.63; 3, 21 mi. E Totolapán, IX.1.63; 7, 12 mi. SE Totolapán, IX.1.63. Puebla: 2, 5 mi. NW Acatlán, IX.11.63; 25, 10 mi. S Acatlán, VIII.21.59, IX.8.59; 1, S Matamoros, IX.25.60 (C. H. Martin); 10, 4 mi. SE Matamoros, IX.3.59; 3, 39 mi. NE Puebla, IX.3.59; 8, 6 mi. N Tehuacán, IX.10.59; 2, 4 mi. S Tehuacán, VIII.22.63. Sonora: 1, 53 mi. S Navojoa, VIII.15.59. Vera Cruz: 2, 9 mi. N Acayucán, IX.13.63.

Paravilla nigrofemorata Hall, new species

Fig. 34

This species is similar to *P. flavipilosa*. It differs from *P. flavipilosa* by its smaller size (8-10 mm) and entirely black femora. The species may be separated by the black femora, entirely pale-yellow pilose body, and near-absence of tomentum on the abdomen of the male. In addition, the wings are at most colored faintly along the costal margin and the first basal cell.

Male. - Black; basal antennal segments, tips of femora,

all of tibiae, basal tarsal segments fulvous.

Front with black hair, absent in middle immediately above antennae, tomentum shining yellow, nearly reaching ocellar tubercle. First antennal segment twice as long as second, broadened apically, black hair on outer surface, mixed yellow and black hair on inner surface; second segment with black hair; third segment elongate conical at base, styliform apical portion about two-thirds longer than basal portion; style short, terminal. Face with black hair and yellow scales. Proboscis not projecting beyond oral margin. Palpus fulvous, with yellow hair. Occiput pale-yellow tomentose; cervical fringe whitish above, pale yellow below.

Mesonotum entirely pale-yellow pilose, tomentum absent, bristles pale. Pleura pale-yellow pilose, tomentum absent. Legs with yellowish hair and scales; all femora with hair; tibial scales pale yellow.

Halter stem testaceous, knob white. Squama light testaceous, with fringe of yellowish scales. Scutellum pale-yellow pilose, tomentum of scattered, short, pale-yellowish hair; bristles pale.

Basicosta with pale-yellow scales and black setulae. Wing hyaline, light yellowish-brown coloring in costal and first basal cells, faint spots of color on r-m crossvein and at base of fourth posterior cell; width of first posterior cell at wing margin a little more than length of r-m crossvein.

Abdominal dorsum pale-yellow pilose, tomentum absent except for pale-yellow appressed hairs on segments five, six, seven, black hair absent. Venter pale-yellow pilose and tomentose.

Genitalia with epandrium about as long as high, lower apical corner rounded, lower margin straight or nearly so; epiphallus straight, tip curved upward at angle of about 45°, apex rounded, lower margin straight; in dorsal view, epiphallus broad, apex rounded, sides slightly concave near middle; aedeagus short, half length of epiphallus, turned downward; basistylus boot-shaped, base rounded, apex bluntly pointed; dististylus narrow, long with shallow hook at apex.

Female. - Front and face yellowish pilose and tomentose; basal antennal segments with yellow hair; thorax, scutellum, and abdomen with abundant pale-yellow tomentum. Otherwise as described for male.

Holotype male. - From Granite Pass, San Bernardino County, California, IV.29.68 (E. I. Schlinger); allotype from Deep Canyon, trail from Hwy. 74, 2690 ft., V.8.73 (A. B. Tabet). Both in collection of California Academy of Sciences.

Paratypes. - Arizona: 1, 10 mi. E Aztec, Yuma County, IV.8.63 (R. C. Dickson); 1, 5 mi. S Oracle Junction, Pinal County, IV.21.63 (M. Wargo). California: 2, topotypic, V.2.73 (A. B. Tabet); 2, Culp Canyon, San Diego County, IV.2.62 (E. I. Schlinger).

Paravilla nigronasica Painter

Fig. 35

Villa (Paravilla) nigronasica Painter, 1933: 14; Painter and Painter, 1965: 439. (Types in University of Minnesota collection.). *Paravilla nigronasica* (Painter), Hull, 1973: 392.

As has been stated by Painter (1933, p. 15), this species is close to *P. xanthina*. It differs from *P. xanthina* by the black pilose front, black antennae, more hyaline wings, and black pilose coxae, as well as the prolonged lower apical corner of the epandrium in the male.

The following redescription of *P. nigronasica* is based on a portion of Painter's type series, in conjunction with additional specimens from within and without the known prior distribution.

Male. - Large size, 10-15 mm in length. Black; middle of side of face, apices of femora, all of tibiae, bases of tarsi fulvous.

Eyes separated at narrowest point by twice width of ocellar tubercle. Front and face with black hair and scattered, shining, golden-yellow scales. First two antennal segments with black hair; first segment cylindrical, not appreciably widened at apex, twice length of second segment; second segment about as wide as long; third segment elongate conical at base, apical styliform portion almost two-thirds longer than basal portion, third segment longer than segments one and two combined. Palpus with black hair. Proboscis not projecting beyond oral margin. Side of occiput with yellow scales, yellow hair at vertex; cervical fringe whitish.

Mesonotum golden-yellow pilose, tomentum absent except along lateral margin above root of wing, bristles fulvous. Pleura yellow pilose and tomentose. Coxae with yellow and black hair and golden-yellow tomentum; black hair on fore and mid femora, black and yellow hair on hind femur, all femora with yellow scales, a few black scales on anterior surfaces of fore and mid femora; tibiae with dark fulvous scales, setulae and bristles black.

Halter stem flavotestaceous, knob whitish. Squama brownish yellow, with fringe of yellow scales. Scutellum golden-yellow pilose and tomentose, tomentum present mainly in middle and along posterior margin; bristles fulvous.

Basicosta with yellow and black scales and black setulae. Wing hyaline, costal cell light yellow; vein at base of second posterior cell somewhat sinuous.

Abdomen golden-yellow pilose and tomentose, latter in bands across bases of segments two to five, black tomentum across posterior margins or more of segments two to five, tomentum on segments six and seven white; side of abdomen golden-yellow-haired. Venter of abdomen golden-yellow pilose and tomentose.

Genitalia with epandrium long, broad at base, narrowed apically, lower apical corner prolonged about twice medial width, lower margin straight; epiphallus bent downward, tip curved upward, apex rounded, lower margin swollen apically; in dorsal view, epiphallus bluntly rounded apically, sides subparallel, slightly concave; aedeagus short, half length of epiphallus, curved downward; basistylus broad basally, bluntly rounded apically; dististylus short, broad, hooked apically.

Female. - Tomentum on mesonotum and thorax golden yellow, dense, terminal abdominal segments with golden-yellow tomentum, that on segment seven sometimes slightly lighter in color; black tomentum on second abdominal segment absent or reduced to a small medial spot. Otherwise as described for male.

Distribution. - Originally described from Nebraska; specimens have been studied from Kansas, Colorado, Utah, Texas, and New Mexico. Almost all of the specimens have been taken at sand dunes or in a sandy habitat.

Paravilla opaca Hall, new species

Fig. 36

This species is similar to *P. separata* and *P. diagonalis*. *P. opaca* can be separated from both species primarily by the long projection of the lower apical corner of the male epandrium. In addition, *P. opaca* has fulvous tomentum on the abdomen along with the black-and-white tomentum, and the antennae and face are black.

Male. - Black; at most oral margin fulvous, femora dusky, tibiae ferruginous or nearly so.

Front and face with black hair and scattered golden-yellow scales. Antennae with black hair, first segment twice length of second, expanded slightly on inner apical margin; second segment a little wider than long; third segment twice length of segments one and two combined, base short conical, styliform apical portion twice as long as conical basal portion. At most, labellae projecting beyond oral margin. Palpus with black hair. Side of occiput with pale-yellow scales, scattered, short, black hair at vertex; cervical fringe pale yellowish.

Disk of mesonotum with short black hair, longer toward scutellum, anterior margin of mesonotum light-yellow pilose, white stripe of hair along lateral margin from humeral callus to post-alar callus, tomentum yellowish, not dense, bristles fulvous. Pleura pale-yellow to white pilose and tomentose, a few black hairs often present on propleuron, lower part of mesopleuron, and sternopleuron. Hair on coxae black and yellow, tomentum yellow; femora with black and pale scales.

Halter stem testaceous, knob creamy yellow. Squama pale yellow, with pale-yellow fringe of scales. Scutellum with black hair, a few white hairs present at sides, tomentum light fulvous; bristles mixed black and fulvous, the latter more numerous.

Basicosta with fulvous scales and black setulae. Basal half of wing infuscated with dark brown, apical half hyaline; color extends diagonally basad from tip of subcostal vein through basal half of discal cell across basal two-thirds of anal cell, and arcs through anterior middle of axillary cell, not reaching posterior margin of wing, both basal cells entirely colored, discal cell colored to r-m crossvein; crossvein at base of second posterior cell bent slightly.

Abdominal dorsum white to pale-yellow pilose, posterior

margins of segments two to five with black hair, hair on last two segments yellow; tomentum light fulvous, white across basal half of second segment and on all of segments six and seven, whitish or pale-yellowish scales across bases of segments three and four, second segment with a short crossband of black scales on posterior half of segment, third and fourth segments with median triangles of black tomentum; lateral margin of abdomen with white hair on segments one to three, black hair on segments four and five, yellow hair on segments six and seven. Venter of abdomen pale-yellow pilose and tomentose.

Genitalia with epandrium narrow, lower apical corner prolonged, prolongation narrow, more than three times longer than median width, lower margin concave; epiphallus humped slightly, broad, tip only slightly curved upward, apex truncate, wrinkled; in dorsal view, epiphallus concave, tip rounded, sides subparallel, converging slightly basad; aedeagus curved downward, less than half length of epiphallus; basistylus narrow, about twice as long as widest point, apex bluntly pointed; diststylus small, hooked apically.

Female. - Face with a little more fulvous ground-color; front with a small patch of black scales between antennae; scutellar bristles black; abdominal segment five almost entirely black tomentose, abdominal dorsum with more obvious bands of pale scales; venter of abdomen with black hair from third segment to apex. Otherwise as described for male.

Variation. - Not all females have the fifth abdominal segment black tomentose nor do they all have the patch of black scales between the antennae. In some specimens, the face is more dusky than black in ground-color and the first antennal segments are dark fulvous.

Holotype male. - From 4 mi. NE and allotype from 21 mi. NE Montemorelos, Nuevo León, Mexico, IV.25.66 (R. H. and E. M. Painter). Both in collection of U.S. National Museum.

Paratypes. - UNITED STATES: Texas: Brown County, 1, Brownwood, Texas, VI.8 (R. H. Painter); Dawson County, 2, 2-10 mi. E Lamesa, IX.13.65 (F. M. and M. C. Hull). MEXICO: Nuevo León, 6, topotypic, collected with types; Baja California del Norte, 3, 21 mi. SE Rosarita, X.9.72 (E. M. Fisher).

Paravilla painterorum Hall, new species

Fig. 37

This species is similar to *P. cinerea*, but it is much smaller and the wings are colored more darkly and extensively. The legs and most of the face are black, the hairs on the body predominantly pale yellow.

Male. - Small size, length 8-11 mm. Black; first antennal segment, side of face partially, tibiae ferruginous.

Front with black hair and pale-yellow scales. First two antennal segments with black hair; first segment one and one-half times longer than second segment, widened apically on inner margin; second segment slightly wider than long; third segment longer than segments one and two combined, base long conical, tapering rather evenly to styliform apex. Face with black hair and scattered pale-yellow scales. Proboscis not projecting beyond oral margin. Palpus with black hair. Side of occiput pale-yellow tomentose, becoming darker yellow to fulvous at vertex; cervical fringe whitish.

Mesonotum yellow pilose, that above root of wing whitish, disk with short black hair, shining yellow hair across posterior margin, tomentum pale yellow, bristles yellow. Pleura pale-yellow pilose and tomentose. Coxae with mixed black and yellow hair, scales yellow; hair on posteroventral surface of femora black, scales on femora and tibiae pale yellow, quite dense.

Halter stem testaceous, knob creamy white. Squama light testaceous, with fringe of whitish scales. Scutellum yellow pilose and tomentose; bristles yellow.

Basicosta with yellow scales and black setulae. Wing with basal half or more infuscated with light brown, apical half or less hyaline; color runs diagonally basad from apex of subcosta to apical fifth of anal cell, then basad along vein 2A, color not reaching posterior margin of wing, base of discal cell colored to r-m crossvein, second basal and anal cells colored a little lighter than first basal cell; vein at base of second posterior cell almost straight.

Base and apex of abdominal dorsum pale-yellow pilose, middle of abdominal segments two to five with scattered, thin, black hair; tomentum yellow, whitish on basal halves of segments two to five, lighter-yellow scales on segments six and seven; lateral margin of abdomen pale-yellow pilose, scattered black hairs present on segments three to five. Venter of abdomen yellow pilose and tomentose.

Genitalia with epandrium longer than median height, lower

apical corner prolonged slightly, apex acuminate, lower margin concave medially, epiphallus straight, tip curved up slightly, apex rounded, lower margin undulated; in dorsal view, tip of epiphallus bluntly pointed, sides subparallel, diverging basally; aedeagus short, less than half length of epiphallus, pointed downward; basistylus rounded at base, apex bluntly pointed; dististylus short, hooked apically.

Female. - Does not differ appreciably from male.

Holotype male and allotype. - From 21 mi. NE Montemoralos, Nuevo León, Mexico, IV.25.68 (R. H. and E. M. Painter). Both in collection of U.S. National Museum.

Paratypes. - MEXICO: Nuevo León: 14, topotypic, collected with types: 1, 16 mi. N Vallecillo, VIII.6.63 (R. H. and E. M. Painter). Zacatecas: 2, 25 mi. NW Zacatecas, IX. 3.62, IX.18.59 (R. H. and E. M. Painter).

Additional Material. - 1 female, Saguache, Saguache County, Colorado, collected in August.

Paravilla palliata (Loew)

Fig. 38

Anthrax palliata Loew, 1869: 20. (Type in Harvard Museum of Comparative Zoology.). *Villa (Paravilla) palliata* (Lw.), Painter, 1933: 10; Painter and Painter, 1965: 439. *Paravilla palliata* (Lw.), 1973: 392.

This is a midwestern species with half-black wings and almost entirely fulvous pilose body, with black hair on the side of the mesonotum between the wing bases and the humeral calli; the legs are entirely fulvous.

Male. - Black; basal antennal segments, face, legs dark fulvous to dark ferruginous, second antennal segment dusky dorsally.

Front and face with black hair and golden-yellow scales, scales not dense. Antennae with black hair; first segment short, expanded apically on inner margin, one and one-half times longer than second segment; second segment about as long as wide; third segment about twice length of first and second segments combined, base short conical, styliform apical portion about one and one-half times longer than basal portion. Proboscis short; at most, labellae projecting beyond tip of oral margin. Palpus with black hair. Occiput with short, scattered, black hair and fulvous scales; cervical fringe pale yellow above, black below.

Mesonotum dark-fulvous to orange pilose and tomentose, latter scattered, hairlike, semi-erect, disk predominantly black-haired, hair longer on humeral callus and along side to wing base, posterior margin of mesonotum with long black hair, bristles fulvous. Pleura fulvous to orange pilose, with black hair on lower half of mesopleuron, on upper portion of sternopleuron, along anterior margin of metapleuron, and on lower portion of propleuron, tomentum, where present, fulvous. Coxae and femora with black hair and fulvous scales, anterior surface of all femora with at least a few black scales; fore tibia with black setulae above, short fulvous hair below; mid and hind tibiae with fulvous scales, a few black scales often present, bristles black.

Halter stem testaceous, knob yellowish. Squama testaceous, with fringe of fulvous scales. Scutellum black pilose, tomentum fulvous, long, semi-erect, hairlike, bristles fulvous, those in middle often black.

Basicosta with fulvous and black scales and black setulae. Basal half of wing dark reddish-brown, apical half hyaline; color runs from apex of subcostal vein in a nearly straight line diagonally basad almost to apex of axillary cell, margin of color sharp, both basal cells colored, base of discal cell colored to r-m crossvein, apical fourth of anal cell hyaline; discal cell swollen apically; vein at base of second posterior cell at most slightly sinuous.

Abdominal dorsum fulvous to orange pilose and tomentose, white scales on segments six and seven, most specimens with a few black scales in middle of second segment (seen under high magnification, x72), short black hair across posterior margins of at least segments two to five; lateral margin of abdomen fulvous pilose from base to apex, lighter on segments six and seven, a few black hairs present on segments three to six. Venter of abdomen fulvous pilose and tomentose, a few black hairs present across posterior margins of most segments.

Genitalia with epandrium almost square, lower apical corner bulged slightly, lower margin straight; epiphallus arched slightly, narrow medially, tip recurved slightly, apex rounded, bulged below, lower margin concave; in dorsal view, tip of epiphallus rounded, sides parallel; aedeagus short, less than half length of epiphallus, directed downward; basistylus narrow, apex bluntly pointed; dististylus short, broad, hooked apically.

Female. - Much like the male except that scales at apex

of abdomen are pale yellow rather than white, and tomentum in general denser and more appressed.

Variation. - In well-preserved specimens, especially females, the dorsum of the abdomen shows small patches of small black scales in the middle of segments two to four. Not all specimens, however, have these black scales present.

Specimens determined as *P. palliata* (and which agree well with the type) have a male genital configuration almost identical to that found in *P. edititoides*. Because of the more orangish-colored hair, the darker brown in the wings, and the presence of hair on the abdominal dorsum, *P. palliata* falls well outside the range of color variations noted in *P. edititoides*. The two species have been taken together in Kansas, but *P. palliata* extends north and eastward, while *P. edititoides* extends south and westward. The latter species is much more commonly collected than is *P. palliata*.

Distribution. - The Catalog of Diptera of North America gives the distribution as Illinois, Colorado, South Dakota, Minnesota, Nebraska, Kansas, and Oklahoma. In addition, I have a specimen from Louisiana. I have not seen the Colorado specimen, but surmise that it is from the eastern side of the Rocky Mountains. These flies are on the wing in August and September.

Paravilla pallida Hall, new species

Fig. 39

This species resembles *P. tricellula*, but may be recognized and separated from *P. tricellula* by the presence of two submarginal cells in the wings, black hair on the femora, almost entirely pale-yellow pilose body, and fulvous legs. In addition, small tufts of black hair are present on the posterior lateral corners of the second and third abdominal segments.

Male. - Length 8-12 mm. Black; first antennal segment, face, legs except coxae, tips of tarsi fulvous.

Front black pilose and fulvous tomentose, scales not reaching to ocellar tubercle. Antennae with black hair; first segment twice as long as second, a little wider on inner apical margin; second segment about twice as wide as long; third segment almost one and one-half times longer than first two segments combined, base long conical, equal in length to apical styliform portion. Face with black hair and scattered fulvous scales. Proboscis not projecting beyond oral margin. Palpus with black hair. Occiput with pale-yellowish scales on side, darker yellow toward vertex; cervical fringe whitish.

Mesonotum pale-yellow pilose, short black hair on disk, longer across posterior margin, whitish hair above root of wing, tomentum yellow, hairlike, not dense, bristles pale. Pleura pale-yellow pilose and tomentose. Coxae with black and yellow hair and yellow tomentum; femora with black hair on posteroventral surfaces and pale-yellow scales, tibiae with pale-yellow scales, setulae and bristles black.

Halter stem testaceous, knob creamy white. Squama flavotestaceous, with fringe of pale-yellow scales. Scutellum with long black hair in middle, yellow hair at sides near base; tomentum and marginal bristles yellow.

Basicosta with yellow scales and black setulae. Wing infuscated with light brown on basal half, apical half hyaline; color runs from tip of subcostal vein diagonally basad across middle of anal cell into basal third of axillary cell, both basal cells infuscated, color in discal cell not reaching r-m crossvein except at vein M_1; vein at base of second posterior cell almost straight.

Abdominal dorsum pale-yellow pilose, black hair present across posterior margins of all segments except first, black hair long and conspicuous, tomentum yellow, black tomentum absent; lateral margin of abdomen pale-yellow pilose, with small tufts of black hair on posterior corners of second and third segments, segments four to seven with mixed black and yellow hair. Venter of abdomen yellow pilose, a few black hairs also present, tomentum golden yellow.

Genitalia with epandrium about one and one-half times longer than median height, lower apical corner rounded, lower margin straight; epiphallus straight, swollen apically, tip curved up, apex rounded, lower margin straight; in dorsal view, epiphallus broad, tip rounded, sides subparallel, convex toward base; basistylus shoe-shaped, base pointed, apex rounded; dististylus small, narrow, not hooked at apex.

Female. - Second antennal segment fulvous, at least below; black hair on abdominal dorsum more pronounced. Otherwise as described for male.

Holotype male and allotype. - From 5 mi. W Pachuca, Hidalgo, Mexico, VIII.14.63 (R. H. and E. M. Painter). Both in collection of U.S. National Museum.

Paratypes. - UNITED STATES: Arizona: Coconino County, 1, S rim Grand Canyon, VII.20.32 (R. H. Painter). Colorado: Denver County, 1, Denver, V.23. Nebraska: Kimball County, 1, 7 mi. W Kimball, VI.27.61 (R. H. Painter). New Mexico: Torrance County, 19, VI.23 (R. H. Painter). Texas: Zapata County, 1, 30 mi. NW Zapata, IV.6.66 (R. H. and E. M. Painter). MEXICO: Chihuahua: 1, 7 mi. SW Chihuahua, IX.6.62 (R. H. and E. M. Painter). Durango: 1, 7 mi. NE Durango, IX.20.50 (R. H. and E. M. Painter). Hidalgo: 6, topotypic, collected with type. Nuevo León: 2, 12 mi. NW Linares, IV.25.66 (R. H. and E. M. Painter); 2, 4 mi. NE Montemorelos, IV.25.66 (R. H. and E. M. Painter). Tamaulipas: 2, 8 mi. N Cuidad Victoria, IV.22.66 (R. H. and E. M. Painter).

Paravilla parasitica Hall, new species

Fig. 40

This species is similar to *P. apicola* in vestiture. The most obvious difference between the two species is the presence of black hair on the side of second abdominal segment in *P. parasitica*; also, the body pile and tomentum are almost entirely pale yellow to light fulvous. The legs are entirely fulvous.

Male. - Black; first antennal segment, face, legs light fulvous.

Front with black hair and pale fulvous scales that reach almost to ocellar tubercle. First antennal segment one and one-half times longer than second segment, cylindrical, not expanded on inner apical margin; second segment at most a little wider than long; both first and second segments with black hair; third segment long conical, with swollen base equal in length to apical styliform portion, segment about three times longer than segments one and two combined. Face with black hair and fulvous scales. Proboscis not projecting beyond oral margin. Palpus with black hair. Side of occiput and vertex with pale-yellow scales, scales with pale-yellow hair; cervical fringe pale yellow.

Disk of mesonotum with short, scattered, black hair, otherwise mesonotum pale-yellow pilose and tomentose, hair above root of wing whitish, bristles yellow. Pleura pale-yellow pilose and tomentose. Coxae with mixed black and light-fulvous hair, scales light fulvous; femora with black hair and setulae and pale-yellow scales; tibiae with pale scales and black setulae.

Halter stem flavotestaceous, knob creamy white. Squama light yellowish, with fringe of pale-yellow scales. Scutellum pale-yellow pilose and tomentose; bristles pale yellow.

Basicosta with light fulvous scales and black setulae. Wing with basal half or less colored with light brown, apical half or more hyaline; color extends diagonally basad in an irregular line from tip of subcostal vein to middle of axillary cell, both basal cells entirely colored, color in discal cell not reaching r-m crossvein except along vein M_1, anal cell colored beyond base of fourth posterior cell, color in axillary cell not reaching hind margin of wing; vein at base of second posterior cell contorted slightly.

Abdominal dorsum pale-yellow pilose and tomentose, a few short black hairs present across posterior margins of segments two to six; lateral margin of abdomen pale-yellow pilose, with scattered black hair on posterior corners of segments two to six. Venter of abdomen yellow pilose and tomentose.

Genitalia with epandrium narrow, twice as long as median height, posterior margin curved basad, lower apical corner rounded, lower margin concave; epandrium thick, straight, tip curved upward slightly, apex rounded, lower margin convex; in dorsal view, epiphallus broad, sides slightly convex, apex rounded; aedeagus long, almost as long as epiphallus, straight, narrow, pointed apically; basistylus long, narrow, much longer than median height; dististylus short, hooked apically.

Female. - Front with mixed black and fulvous hair; face with a stripe of fulvous hair below antennae toward eye margin. Otherwise as described for male.

Holotype male and allotype. - From Paulden, Yavapai County, Arizona, VI.4.55 (D. E. and L. M. Johnson). Both in collection of University of Arizona.

Paratypes. - UNITED STATES: Arizona: Pima County, 1, 10 mi. W Tucson, IV.1.61 (R. H. and E. M. Painter); Pinal County, 34, 6 mi. E. Sacaton, IV.24.75 (M. Linsley, F. Werner), flying over a gregarious bee *Diadasia diminuta* (Cress.) nest in a dirt road; Yavapai County, 7, topotypic, collected with types. California: Riverside County, 1, Palo Verde,

IV.22.35 (F. H. Parker). New Mexico: Eddy County, 5 mi. N Carlsbad, IX.18.56 (J. W. MacSwain).

Paravilla parvula Hall, new species

Fig. 42

This is a small (6-10 mm in length) species separated by the following characters: basal half or less of the wings colored; body pile yellow, with tufts of black hair on the sides of abdominal segments two to four and with an enlarged spot or crossband of black tomentum on the posterior margin of the second abdominal segment; femora black.

Male. - Black; first antennal segment, at least side of face, tips of femora, all of tibiae fulvous.

Front with black hair and narrow fulvous scales not reaching to ocellar tubercle. Antennae with black hair; first segment cylindrical, not expanded apically, one and one-half times longer than second segment; second segment a little wider than long; third segment twice as long as segments one and two combined, basal portion short conical, apical styliform portion twice length of basal portion. Face with black hair and fulvous scales. At most, labellae projecting beyond oral margin. Palpus with black hair. Occiput with yellowish tomentum and scattered, short, black hair; cervical fringe yellowish on upper half, black on lower half.

Mesonotum with short, scattered, black hair on disk, yellow hair elsewhere, that above root of wing whitish; prealar bristles black, one or two yellow bristles also present, post-alar bristles yellow. Pleura yellowish pilose and tomentose, with small patch of black hair on mesopleuron. Fore and mid coxae with long black hair, hind coxae with fulvous and black hair, tomentum on all coxae fulvous; femora with black hair on posteroventral surfaces, fulvous tomentose, tibiae with fulvous scales, setulae and bristles black.

Halter stem brown, knob creamy white. Squama yellowish, with fringe of yellow scales. Scutellum yellow pilose and tomentose; bristles yellow.

Basicosta with yellow scales and black setulae. Basal half of wing infuscated, apical half hyaline; color runs diagonally basad from tip of subcostal cell to basal half of axillary cell, margin of color sharp, both basal cells entirely colored, color in discal cell not reaching r-m crossvein, apical third of anal cell hyaline, no darker spots of color on crossveins; vein at base of second posterior cell bent slightly.

Abdominal dorsum yellowish pilose and tomentose, black hair in middle of posterior margins on segments two to five, that on segments six and seven reaching sides of segments; black tomentose crossband on posterior half of second segment not reaching lateral margin of abdomen, black tomentum on segments three to five (when present) reduced to small median spots; lateral margin of abdomen yellowish pilose, with definite tufts of black hair on the posterior corners of segments two to four, that on rest of lateral margin scattered. Venter of abdomen fulvous pilose and tomentose, with a few scattered black hairs present on apical half of venter.

Genitalia with epandrium, in lateral view, short, about

half as long as high, lower apical corner rounded, lower margin concave; in lateral view, epiphallus straight, rounded apically, tip curved up, lower margin concave slightly in middle; in dorsal view, tip of epiphallus bluntly rounded, sides concave before middle; aedeagus short, about half length of epiphallus, directed downward; basistylus narrow, pointed apically; dististylus narrow, long, hooked apically.

Female. - As described for male except that tomentum on front reaches ocellar tubercle.

Variation. - The venter of abdomen may be almost entirely black-haired or with only one or two black hairs present. Larger specimens often have a complete crossband of black tomentum on the third abdominal segment as well as on the second segment; the black tomentum is often in the form of small, median, triangular spots on segments two and three. The specimens from the more northern latitudes are somewhat lighter in hair coloring than are those from the south.

Holotype male and allotype. - From Estancia, Torrance County, New Mexico, VIII.12.31 (R. H. Painter). Both in collection of U. S. National Museum.

Paratypes. - UNITED STATES: Arizona: Apache County, 2, 5 mi. SE Round Rock, VIII.28.67 (J. C. Hall); Cochise County, 9, Chiricahua Mtns., IX.15.35 (D. K. Duncan, F. H. Parker); 187, Southwest Research Station, Portal, IX.5-25.65, IX.9-X.2.66 (C. W. Sabrosky, P. H. Arnaud, Jr.); 6, Texas Canyon, Chiricahua Mtns., IX.6.66 (L. F. LaPré); Coconino County, 1, nr. Flagstaff, Ft. Valley, IX.1.69 (W. L. Nutting); 3, Flagstaff, VIII.21.39, X.2.57, VIII.23.67 (F. E. Van Dyke,

J. C. Hall, E. M. and R. H. Painter); 1, 14.2 mi. N Flagstaff, VIII.23.67 (J. C. Hall); 10, Grand Canyon, VIII.1971 (F. M. Hull); Graham County, 2, Stockton Pass, 7.3 mi. E Bonita, IX.3.66 (K. W. Brown); 3, White Mtns., IX.21.36 (F. H. Parker); Santa Cruz County, 1, 10 mi. N Lochial, IX. 17.72 (G. R. Ballmer); 1, Madera Cyn. Sta., Rita Mtns., IX. 28.56 (F. G. Werner); 1, 6 mi. E Patagonia, IX.17.71 (J. L. Bath); 4, Snowflake, VIII.20.57 (E. M. and R. H. Painter); 2, Sonoita, IX.19.35 (R. H. Parker). Colorado: Archuleta County, 9, 3 mi. W Pagosa Springs, VIII.25.65 (R. H. and E. M. Painter); Cheyenne County, 2, Kit Carson, VIII.16.30 (D. A. Wilbur); Gunnison County, 1, Paonia, VIII.13.46 (R. H. Painter); Huerfano County, 1, 2 mi. W Walsenburg, VIII.24.65 (R. H. and E. M. Painter); 4, VIII.14.28 (R. H. Painter); Kiowa County, 2, VIII.19 (R. H. Painter); Mesa County, 1, 4.5 mi. S Mesa, VIII.25.67 (J. C. Hall); Montezuma County, 1, Mesa Verde, VIII.23.34 (F. E. Lutz); Phillips County, 2, 2 mi. W Holyoke, IX.2.64 (R. H. and E. M. Painter). Idaho: Oneida County, 2, 5 mi. NW Holbrook, VIII.26-28.72 (Hanson and Knowlton), malaise trap. Kansas: 1, West Kansas, VIII (D. A. Wilbur). New Mexico: Bernarillo County, 2, Cline Corners, E Albuquerque, VIII.25.40 (C. W. Sabrosky); 1, 7 mi. W Albuquerque, IX.7.64 (F. M. and M. C. Hull); 2, 17 mi. W Albuquerque, IX.16-17.68 (P. H. Arnaud, Jr.); Catron County, 1, Quemado, VII.11.30 (T. F. Winburn, R. H. Painter); Colfax County, 1, Koehler (W. R. Walton) n.d.; Grand County, 2, 22 mi. SW Silver City, VIII.27.67 (W. J. Hanson); Guadalupe County, 20, Hwy. 64, 14 mi. W Santa Rosa, IX.6.64

(F. M. and M. C. Hull); 9, Hwy. 64, 26 mi. W Santa Rosa, IX.7.64 (F. M. and M. C. Hull); 58, 60 mi. W Santa Rosa, IX.7.64 (F. M. and M. C. Hull); Lincoln County, 10, 12 mi. SW Capitan, IX.12.62 (R. H. and E. M. Painter); Otero County, 1, 7 mi. N Ruidoso, IX.12.62 (R. H. and E. M. Painter); San Miguel County, 2, nr. Las Vegas Hot Springs, IX (F. H. Snow); Socorro County, 4, Claunch, IV.11.68 (P. H. Timberlake); 1, Gran Quivera, VIII.11.31 (R. H. Painter); Torrance County, 27, topotypic, collected with type; 3, Tajique, VIII.12.31 (R. H. Painter); 2, Tajique, VII.25 (C. H. Martin); 13, New Mexico (F. H. Snow). Texas: Jeff Davis County, 2, Davis Mtns., X.17.57 (E. M. and R. H. Painter). Utah: San Juan County, 2, 5 mi. S Monticello, VIII.29.65 (R. H. and E. M. Painter). MEXICO: All, unless otherwise noted, collected by R. H. and E. M. Painter. Chihuahua: 1, 47 mi. SW Chihuahua, IX.6.62. Durango: 15, 31 mi. W Durango, IX.19-20.62; 26, 9 mi. NE, 28 mi. N, 39 mi. W, 45 mi. W Durango, IX.19-20.59. Guanajuato: 9, 3-4 mi. N León, IX.19.59, IX.23.63; 1, Comanjilla, IX.23.63. Hidalgo: 1, 18 mi. SE Actopán, VIII.13.63; 3, 5 mi. W Pachuca, VIII.14.63. Jalisco: 1, 8 mi. W Lagos de Moreno, VIII.29.62. Mexico: 5, 1.4 mi. S Chapingo, IX.25.63; 9, 1 mi. NW, 4 mi. NE Toluca, VIII.19-20.62, IX.19.63. Michoacán: 9, 3-8 mi. W Morelin, IX.20.63. Nuevo León: 3, 7 mi. W Iturbide, IX.26.63. San Luis Potosí: 3, 6 mi. W El Salto, IX.4.62; 1, 3 mi. N Matehuala, IX.25.63. Sonora: 7, Yecora, X.10.59 (W. W. Gibson). Zacatecas: 1, 19 mi. NW Fresnillo, IX.3.62; 1, 25 mi. NW Zacatecas, IX.18.59.

Paravilla perplexa (Coquillett)

Fig. 41

Anthrax perplexa Coquillett, 1887: 176. (Type in U.S. National Museum.). *Villa (Paravilla) perplexa* (Coq.), Painter and Painter, 1965: 439. *Paravilla perplexa* (Coq.), Hull, 1973: 392.

This species is perplexing in that it shares many characteristics with *P. vigilans*, *P. syrtis*, and *P. lacunaris* and is difficult to include in a key or to diagnose. At the present time, *P. perplexa* is known from the type series only. This species also resembles *Chrysanthrax* species. *P. perplexa* is small (5-7 mm in length), with half-colored wings, white pilose body, and legs that are fulvous except for the dusky fore femora.

Male. - Black; lower fourth of front, first antennal segment, face, except sometimes middle above, legs except fore femora, terminal abdominal segments fulvous.

Front with black hair and pale-fulvous scales, dense above and between antennae. First antennal segment short, twice as long as second segment, expanded on inner apical margin; second segment much broader than long, both first and second segments with black hair; third segment about three times longer than first two segments combined, base short conical, apical styliform portion about twice as long as basal portion. Face with black hair and light-fulvous scales. Labellae projecting beyond tip of oral margin. Palpus with yellow hair. Side of occiput white tomentose, becoming yellow at vertex; cervical fringe yellow.

Mesonotum pale-yellowish pilose and yellowish tomentose,

tomentum dense, disk with short black hair not reaching the posterior margin of the mesonotum, stripe of hair along lateral margin above root of wing white, bristles pale yellow. Pleura pale-yellow pilose and tomentose. Coxae yellow pilose and tomentose, each coxa with a few black hairs present; fore and middle femora with black and/or yellow hair on posteroventral surface, hair absent on hind femur, scales on femora pale yellow.

Halter stem fulvous, knob creamy yellow. Squama whitish, with fringe of whitish scales. Scutellum yellow pilose, pale-yellow tomentose; bristles yellow.

Basicosta with yellow scales and black setulae. Wing with basal half or less infuscated; color runs from apex of subcosta diagonally basad in a jagged line to apical fourth of anal cell, extending into axillary cell along vein 2A only, color in most specimens lighter toward base of wing, color in base of discal cell reaching r-m crossvein, without spots of color in hyaline portion of wing; vein at base of second posterior cell undulated, not sinuous.

Abdomen white pilose on three basal segments, gradually shading to yellow on apical segments; tomentum pale yellow and dense, median, laterally expanded spots of black scales on posterior margins of segments two and three and sometimes four, two specimens in series without black scales; lateral margin of abdomen entirely white to pale-yellowish pilose. Venter of abdomen with yellow hair and whitish scales.

Genitalia with epandrium higher than long, lower apical corner acute, expanded slightly, lower margin straight; epiphallus somewhat humped, tip curved upward, apex rounded, wrinkled, lower margin straight; in dorsal view, epiphallus rounded apically, sides below apex strongly concave, widened basally; aedeagus short, bent downward; basistylus long, narrow, apex pointed; dististylus narrow, long, hooked apically.

Female. - Black hair on abdominal incisures and on legs; otherwise as described for the male.

Distribution. - Known only from Los Angeles County, California.

Coquillett mentioned that 36 specimens were in the type series. Only six remain in the National Collection.

Paravilla separata (Walker)

Fig. 43

Anthrax separata Walker, 1852: 177. (Type in British Museum, Natural History collection.). *Anthrax nemakagonensis* Graenicher, 1910: 26; Painter and Painter, 1965: 439. *Villa (Paravilla) separata* (Wlk.), Painter and Painter, 1962: 124, 1965: 439. *Paravilla separata* (Wlk.), Hull, 1973: 392.

The half-black wings of this eastern species are reminiscent of *Hemipenthes* species. The closest apparent relative to *P. separata* is *P. palliata*; the two may be separated by the black scutellar bristles and black tomentose crossbands on abdominal segments two to four in *P. separata*.

Male. - Black; lower fourth of front, at least antennal segment, face, apical half or more of fore femora, rest of all legs fulvous to castaneous.

Front with black hair and testaceous scale, the latter not approaching ocellar tubercle. First antennal segment two and one-half times longer than second segment, expanded on inner apical margin; second segment slightly wider than long, both segments with black hair; third segment longer than first two segments combined, base long conical apical styliform portion equal in length to basal portion. Face with black hair and yellow scales. Palpus with black and yellowish hair. Side and vertex of occiput with pale-yellow scales, a few black hairs at vertex; cervical fringe yellow on upper half, black on lower half.

Mesonotum with short black hair medially, longer toward scutellum, collar and side in front of wing base yellow, whitish appressed hair above root of wing, tomentum yellow, golden in some specimens, bristles yellow, a small clump of shorter bristles in front of wing base and on post-alar calli. Pleura pale-yellow pilose and tomentose, hair on upper part of mesopleuron whitish, a few black hairs present on propleuron and lower part of mesopleuron, a few black hairs also present between fore and mid coxae. Coxae with black hair and yellow scales; fore and mid femora with black hair, absent on hind femur, all femora with yellowish scales, darker scales sometimes present on anterior and ventral surfaces; tibiae with yellowish scales, bristles black.

Halter stem testaceous, knob whitish yellow. Squama flavotestaceous, with fringe of yellow scales. Scutellum with black hair and bristles, tomentum yellow to golden.

Basicosta with yellow scales and black setulae. Basal half or less of wing dark brown, apical half or more hyaline; color runs from tip of subcosta diagonally basad in an irregular line to apical fourth of anal cell, thence to posterior margin at middle of axillary cell, color at base of discal cell reaching beyond r-m crossvein along vein M_1; knee of vein R_4 usually with a short stump of vein; vein at base of second posterior cell undulated, not sinuous.

Abdominal dorsum pale-yellow pilose, black hair at least in middle along posterior margins of segments one to five; tomentum yellow, black tomentose crossbands on posterior portions of segments two to four, that on four occupying almost entire segment, in partially abraded specimens black scales also visible on segments five and six; side of segments one and two with pale-yellow hair, that at base of third segment a little darker yellow, tufts of black hair at posterior corners of segments two and three, hair on segments four and five almost entirely black, yellow hair at sides of segments six and seven. Venter of abdomen yellow pilose and tomentose, with a few scattered black hairs present on apical segments.

Genitalia with epandrium rectangular, lower apical corner projecting slightly, lower margin straight or at most slightly concave; epiphallus thick, humped slightly, tip curved upward, apex rounded, wrinkled, lower margin straight; in dorsal view, epiphallus rounded apically, sides slightly concave near middle, widened basally; aedeagus short, about half length of epiphallus, pointed downward; basistylus boat-shaped, apex blunt, rounded; dististylus short, broad, apex with a small hook.

Female. - As described for male except for denser tomentum and more prevalent black hair on the thorax and abdomen.

Variation. - Some specimens have the first antennal segment, face and almost all of the legs entirely black; the darker specimens also have more black hair and scales than usual.

Distribution. - Found from Manitoba, Canada, south through South Dakota and Kansas, to the Gulf states, including Florida. In the east, it is found in Michigan, Minnesota, Wisconsin, and Ohio. The Catalog of Diptera of North America lists *P. separata* from Oregon. I have not seen specimens west of the Rocky Mountains.

Paravilla spaldingi (Painter)

Fig. 44

Villa (Paravilla) spaldingi Painter, 1933: 13, 1965: 439. (Type in U.S. National Museum.). *Paravilla spaldingi* (Paint.), Hull, 1973: 392.

The half-black wings, black pile with bright-fulvous hair at the base of the abdomen and on the thoracic collar, and white scales at the apex of the abdomen make this the most distinctive species of *Paravilla*.

Male. - Black; oral margin, side of face fulvous.

Front and face with black hair and fulvous scales, scales reaching ocellar tubercle. First antennal segment twice length of second segment, at most expanded slightly on inner apical margin; second segment as wide as long, both first and second segments with black hair; third segment longer than first two segments combined, base long conical, apical styliform portion equal in length to basal portion. At most, tip of labellae projecting beyond oral margin. Palpus with black hair. Side of occiput yellow tomentose, vertex with short black hair; cervical fringe whitish on upper half, black on lower half.

Mesonotum black pilose, yellow to fulvous hair on collar; tomentum yellow to fulvous, long, hairlike, more dense and longer above root of wing; pre-alar bristles fulvous and black, post-alar bristles black. Pleura black pilose and tomentose, fulvous hair on upper part of propleuron and mesopleuron, metapleural tuft of hair mixed black and fulvous. Legs with black hair and scales.

Halter stem dark brown, knob creamy white. Squama brown, with fringe of fulvous scales. Scutellum black pilose and tomentose, some scattered fulvous tomentum present at sides near base; bristles black.

Basicosta with black scales and setulae. Wing with basal two-thirds black, rest hyaline, outline of color indistinct; color runs from apex of subcosta diagonally basad to tip of anal cell, axillary cell entirely colored, all but apical third of discal cell colored; vein at base of second posterior cell undulated.

Abdominal dorsum black pilose, first and basal half of second segments with yellow to bright fulvous hair, segments six and seven with white hair, tomentum black, that on sixth and seventh segments white; lateral margin of abdomen with yellow to bright-fulvous hair on segment one and basal half of two, apical half of second segment, all of third, fourth, and fifth segments with black hair, white hair on segments six and seven. Venter of abdomen black pilose and tomentose.

Genitalia with epandrium, except for petiole, almost square, lower apical corner prolonged for a distance equal to width of prolongation, lower margin concave; epiphallus thick, straight, tip somewhat curved up, apex rounded, wrinkled, lower margin with medial bulge; in dorsal view, epiphallus broad, side subparallel, apex rounded; aedeagus short, curved downward; basistylus long, apex bluntly rounded; dististylus long, hooked strongly at apex.

Female. - Except for denser tomentum, female does not differ from male.

Variation. - A few specimens lack white hair or scales on the apex of the abdomen. The bright-fulvous pile varies from yellow to bright fulvous.

Distribution. - West of the Rocky Mountains from Utah to California, north to Oregon, thence south as far as Jalisco, Mexico.

Paravilla splendida Hall, new species

Fig. 45

This species resembles *P. lacunaris*, the primary difference being in the darker coloration of the body hairs and wing infuscation in *P. splendida*. In *P. splendida*, there is a spot of color at the base of the second posterior cell and sometimes at the base of vein R_4, the spots of color often quite light. This species is found in the chaparral zones of southern California and Baja California del Norte, Mexico, from the coast inland, while *P. lacunaris* is found in the lower Sonoran or desert areas of southern California.

Male. - Black; first antennal segment, face, legs except tips of tarsi, terminal abdominal segment fulvous.

Front with black hair and light-fulvous scales, scales reaching to ocellar tubercle. Antennae short, first segment one and one-half times length of second segment, expanded on inner apical margin, with black hair; second segment wider than long, with black hair; third segment short, at most only slightly longer than segments one and two combined, base long conical, apical styliform portion equal in length to basal portion. Face with black hair and pale-fulvous scales, a few fulvous hairs below antennae. Proboscis short; at most, labellae projecting beyond oral margin. Palpus with black hair. Side of occiput white tomentose, becoming yellow at vertex; cervical fringe pale yellow.

Mesonotum pale-yellow pilose, black hair on disk, white hair along lateral margin above root of wing, tomentum pale fulvous, bristles fulvous. Pleura pale-yellow to white pilose and tomentose. Coxae pale-yellow pilose, white tomentose; fore and mid femora with yellow hair on posteroventral surfaces, hair absent on hind femur, scales on femora and tibiae fulvous, small patch of black scales at apex of femora on anterior surfaces, bristles black.

Halter stem yellow, knob white. Squama flavotestaceous, with fringe of yellow scales. Scutellum black-haired, tomentum light fulvous, whitish across base; bristles fulvous.

Basicosta with fulvous scales and black setulae. Wing infuscated on basal half or more; color extends diagonally basad from tip of subcostal cell to apex of anal cell, axillary cell colored in middle only, color in discal cell not reaching r-m crossvein except as a broad border to vein M_1, spot of light color on vein at base of second posterior cell, spot of faint color at base of vein R_4, second basal cell subhyaline, color in first posterior cell extending almost to end of discal cell.

Abdominal dorsum pale-yellow pilose, at most a few short, inconspicuous black hairs present in middle of posterior margins of segments two through four; tomentum pale yellow across bases of segments two, three, and four, darker-yellow to light-fulvous scales on side of segment two, across most of posterior margins of segments three and four, and on all of segments five through seven, black tomentose triangles in middle of segments two and three; lateral margin of abdomen entirely pale-yellow pilose. Venter of abdomen pale-yellow to whitish pilose and tomentose.

Genitalia with epandrium about twice as long as median height, lower apical corner bluntly pointed, lower margin shallowly concave medially; epiphallus thickset, humped, tip curved upward, apex rounded, wrinkled, lower margin almost straight, side above with microsetae; in dorsal view, epiphallus narrow, apex rounded, sides concave below apex; aedeagus short, slightly more than half length of epiphallus, pointed downward; basistylus boat-shaped, apex rounded; dististylus short, hooked apically.

Female. - Does not differ appreciably from the male.

Variation. - In many specimens the abdomen is entirely pale-yellow tomentose, with only a small spot of black scales in the middle of the second segment; in others, the black tomentum forms an almost complete crossband on the second segment.

Holotype male and allotype. - From 4 mi. W Cajon Junction, San Bernardino County, California, VII.4.58 (E. I. Schlinger). Both in collection of U.S. National Museum.

Paratypes. - UNITED STATES: California: Kern County, 3, Walker Pass, VII.8.57 (J. C. Hall); Los Angeles County, 2, Tanbark Flat, VII.10.50 (J. C. Hall); Riverside County, 1, Millard Canyon, V.10.63 (E. I. Schlinger); San Bernardino County, 11, topotypic, collected with type (E. I. Schlinger, J. C. Hall); 5, Cajon Pass, VII.24.30, VII.17.58 (R. H. Painter, T. F. Winburn, H. R. Moffitt, J. C. Hall); 1, San Bernardino, VII.23.30 (T. F. Winburn, R. H. Painter); San Diego County, 2, Anza, VII.5.54 (H. R. Moffitt, P. D. Hurd); 41, 1 to 5 mi. SE Del Mar, VII.17.63 (E. I. Schlinger, J. C. Hall); 2, La Jolla, VI.25.63 (E. I. Schlinger). MEXICO: Baja California del Norte: 1, 4 mi. S La Rumorosa, VII.8.67 (M. E. Irwin).

Paravilla syrtis (Coquillett)

Fig. 46

Anthrax syrtis Coquillett, 1887: 173. (Type in U.S. National Museum.). *Villa (Paravilla) syrtis* (Coq.), Painter and Painter, 1965: 439. *Paravilla syrtis* (Coq.), Hull, 1973: 392.

This is a widespread polymorphic, bivoltine species. It is among the commonest of the genus. Because of the polymorphism, *P. syrtis* is difficult to diagnose. Perhaps the most reliable characteristic, other than those of the male genitalia, is the presence on the abdominal dorsum of fulvous scales that normally separate the white and black scales on the second and third segments. The lower apical corner of the epandrium in the male is rounded or pointed.

In the spring and early summer ("spring forms") two types of males are found: (1) those with definite tufts of black hair at the sides of second abdominal segment; and (2) those without black hair on the sides of the abdomen anterior to the fourth segment. Individuals can be found not only of these types, but also of intermediates, i.e., those with only a few black hairs present laterally. Both sexes of the spring-early summer forms almost always have black tomentum on the second abdominal segment, in the form either of a crossband or a median triangle. The black tomentum is bounded basally by fulvous scales, followed by a band of white scales across the base of the second segment; this pattern is also frequently found on the third segment.

In the late summer or fall form of *P. syrtis*, the black tomentum on the second abdominal segment is usually absent. There are almost always some black hairs present at the sides of the abdomen. The femora are frequently dusky in color, and the face and antennae are black in ground-color.

The wing coloring in both forms of *P. syrtis* also exhibits slight variation, in that sometimes a vague spot of color surrounds the crossvein at the base of the second posterior cell.

In the following redescription, the more common spring-early summer form is described, followed by a shorter comparative description of the late summer-fall form. It must be borne in mind that the descriptions are drawn from the most typical forms and that intermediates are frequently encountered.

Male (spring-early summer). - Black; first two antennal segments, face, legs except tarsi, terminal abdominal segments fulvous to ferruginous.

Front with black hair and testaceous scales, latter not reaching ocellar tubercle. First antennal segment twice length of second segment, expanded on inner apical margin, with black hair; second segment rounded, slightly wider than long, with short black hair; third segment longer than first two segments combined, base long conical, apical styliform portion three-fourths length of conical basal portion. Face with black hair and testaceous scales, a small cluster of fulvous hairs below base of first antennal segemnt. Palpus with black hair. Proboscis not projecting. Side of occiput with white scales, becoming yellowish toward vertex; whitish hair at vertex short and inconspicuous; cervical fringe of whitish to pale-yellow hair.

Mesonotum with short black hair on disk, yellow hair on anterior margin of mesonotum, posterior margin with mixed black and yellow hair, lateral margin with white hair somewhat appressed; tomentum testaceous, most prevalent on posterior two-thirds of notum, anterior third of notum with silky, white, hairlike tomentum, most prevalent in middle, bristles fulvous. Pleura pale-yellow to white pilose, patch of dense hairlike tomentum on upper part of sternopleuron and above hind coxa. Coxae with mixed black and fulvous hair, tomentum pale testaceous; fore femur with black hair, mid and hind femora with mixed black and yellow hair on posteroventral surfaces, pale scales on femora and tibiae, bristles and setulae black.

Halter stem flavotestaceous, knob creamy white. Squama testaceous, with fringe of pale-yellow scales. Scutellum with black hair, white hair at sides, tomentum pale yellow, white across basal half; bristles fulvous.

Basicosta with fulvous scales and black setulae. Wing infuscated with brown on basal half or less; color runs diagonally basad in an irregular line from tip of subcostal vein to base of third and fourth posterior cells, then basad almost in a straight line across anal cell, then curving through basal half of axillary cell, color not reaching hind margin of wing, basal halves of both second basal and anal cells subhyaline, color in discal cell reaching r-m crossvein only as a border along vein M_1, faint or indistinct spots of color often present at base on vein R_4 and at base of second posterior cell (best seen in indirect light); crossvein at base of second posterior cell almost straight.

Abdominal dorsum with pale-yellow hair, white hair in middle of segments, posterior margins of all segments except the first, sixth, and seventh with black hair, hair at apex of abdomen yellow, tomentum white across basal halves of segments two and three, black tomentose crossbands across apical halves of segments two and three separated from white tomentum by a narrow line of fulvous scales, tomentum elsewhere from pale yellowish to fulvous; lateral margin of abdomen with white hair on segments one to five, that on sixth and seventh segments pale yellowish, black hair on posterior halves of segments two through five. Venter of abdomen pale-yellow to white-haired; tomentum on first segment and basal half of second segment, all of sixth and seventh segments, and at sides of segments three to five white, rest fulvous.

Genitalia with epandrium narrow, about twice as long as high, lower apical corner rounded, lower margin concave medially; epiphallus humped slightly, tip curved upward, apex rounded, lower margin more or less straight; in dorsal view, epiphallus broad, apex narrowed, rounded, sides subparallel, slightly concave below middle; aedeagus half length of epiphallus, curved downward slightly; basistylus broadest at base, narrowed to acuminate apex; dististylus small, hooked apically.

Female. - Third antennal segment with styliform portion equal in length to conical basal portion; abdomen in general more white-haired than pale-yellow-haired; black hair across posterior margins of segments two to seven, tomentum denser; venter with scattered black hairs. Otherwise as described for male.

Male and female (late summer-fall form). - As described for spring form except antennae and face black in ground-color; at least fore femur dusky, usually all femora with some dusky color, abdominal ground-color black; tomentum on thorax and abdomen predominantly white, middle of second segment and posterior margins of segments three to five with pale yellow tomentum; black scales absent on abdominal dorsum; lateral margin of abdomen with black hair from posterior corner of second segment to apex; lower apical corner of epandrium pointed.

Variation. - Almost every character exhibits some variation, not only between the sexes but also between the spring and fall forms. Male may or may not have black hair at the side of the second abdominal segment in the spring form; the fall form may have black tomentum on the second abdominal segment. Variations are continuous and are found within single populations.

Distribution. - Found throughout the western portion of North America, from Canada to northern Mexico. It is a common spring species in the Southwest, occurring throughout the lower Sonoran.

Paravilla texana Hall, new species

Fig. 47

This species is similar to *P. albicera*, differing mainly in the black-and-yellowish fasciate abdomen. Wing are colored lightly on the basal half with light brown, or may be hyaline with faint spots of color on r-m and posterior crossvein; femora are dusky but not definitely black.

Male. - Black; lower half of front, first two antennal segments, face, terminal abdominal segments fulvous, legs dusky fulvous.

Front with yellow hair and pale-yellow scales, black

hair on ocellar tubercle. First two antennal segments with mixed black and yellow hair; first segment three times longer than second segment, cylindrical, not expanded apically; second segment one and one-half times wider than long; third segment longer than two basal segments together, base long conical, apical styliform portion equal to broadened basal portion. Face with black hair and pale-yellow scales, a few scattered yellow hairs present on sides below. Labellae projecting beyond tip of oral margin. Palpus with black hair. Scales on side of occiput whitish, whitish hairs at vertex; cervical fringe pale yellow to whitish.

Mesonotum with short black hair on disk, pale-yellow hair on anterior margin and lateral margin in front of wing base, hair above base of wing and across posterior margin more whitish than yellowish, tomentum dense, pale yellowish, bristles yellow. Pleura pale-yellow to white pilose and tomentose. Hair on coxae fulvous, scales yellowish; fore and mid femora with black hair, hair absent on hind femur, scales on femora flavotestaceous, bristles black.

Halter stem flavotestaceous, knob white. Squama flavotestaceous, with fringe of pale-yellow scales. Scutellum with short black hair in middle, pale-yellow hair elsewhere; tomentum pale yellow, dense; bristles, yellow.

Basicosta with pale-yellow and whitish scales and black setulae. Wing hyaline, at most lightly colored with brown on basal half, outline of color indistinct, slightly darker spots of color on r-m crossvein and at bases of third and fourth posterior cells, second basal and anal cells subhyaline, the latter almost to apex; vein at base of second posterior cell sinuous.

Abdominal dorsum white to pale-yellow pilose, pile short, short black hair across posterior margins of segments two to five; tomentum pale yellow, white on basal half of second and third segments and all of sixth and seventh segments; crossbands of black tomentum on posterior halves of segments two to four, none of the crossbands reaching the side of the abdomen; lateral margin of abdomen pale-yellow pilose, a few black hairs present on segments four and five. Venter of abdomen white pilose and tomentose on segments one to three, segments four through seven with yellow hair and scales.

Genitalia with epandrium long and comparatively narrow, length more than twice median height, lower apical corner bluntly acuminate, lower margin slightly convex; epiphallus at most humped slightly, tip recurved slightly, apex thickened below, rounded, lower margin straight; in dorsal view, tip of epiphallus broad, rounded, sides narrowed basally; aedeagus half length of epiphallus, pointed downward; basistylus short, broad basally, apex acuminate; dististylus short, narrow, hooked apically.

Female. - Tomentum on abdominal segments two to five pale yellow except for black crossbands, black hair on abdomen much denser, fifth segment with crossband of black tomentum; wing usually slightly darker; venter of abdomen yellowish pilose and tomentose, with a few black hairs near apex. Otherwise as described for male.

Variation. - In some specimens the middle of the scutellum is reddish brown; others have black scales on the anterior surfaces of the femora. A few males have the abdomen patterned as in the females, that is, with a black tomentose crossband on the posterior half of the fifth segment. The hair on the body varies from pale yellow to almost fulvous.

Holotype male and allotype. - From Crane County, Texas, VI.21.58 (R. H. and E. M. Painter). Both in collection of U.S. National Museum.

Paratypes. - UNITED STATES: Colorado: Prowers County, 1, Holly, VII.30.57 (N. Marston). Kansas: Clark County, 1, no data, (F. H. Snow); Kiowa County, 3, VII.17.58 (R. H. Painter); Meade County, 4, VII.18.58 (R. H. Painter); Morton County, 1, VI.1905 (F. H. Snow); Pratt County, 2, VII.19.58 (R. H. Painter); Scott County, 1, VI.14.39 (R. H. Painter); Seward County, 5, VI.1929 (R. H. and E. M. Painter); Trego County, 2, VII.12.12 (F. X. Williams). Nevada: Nye County, 1, Mercury, Nevada Test Site, VI.11.61. New Mexico: Dona Ana County, 3, Hatch, VII.16.52 (R. H. and L. D. Beamer, W. LaBerge); 2, Las Cruces, VI.1910 (D. W. Coquillett); 2, Mesilla Park, VII.15.52 (R. H. Painter, L. D. Beamer, W. LaBerge); Guadalupe County, 1, Vaughan, VII.30.52 (R. H. and L. D. Beamer, C. Liang, W. LaBerge); Hidalgo County, 3, Rodeo, VIII.3.59, VIII.19.59 (H. E. Evans); Roosevelt County, 1, Portales, VII.16.36 (M. B. Jackson). Texas: Atacosa County, 1, Lytle, VI.1.51 (P. D. Hurd); Bexar County, 1, San Antonio, VII.12.31 (G. P. Engelhardt); Brown County, 1, Brownwood, VIII.28.31 (R. H. Painter). Cameron County, 1, Brownsville, no data (Townsend); Crane County, 7, topotypic, collected with type; Goliad County, 2, Goliad, V.8.53 (L. D. Beamer); Grath County, 1, VI.6.69 (R. H. and E. M. Painter); Hemphill County, 3, Canadian, VII.24.52 (R. H. and L. D. Beamer, W. LaBerge); Klegberg County, 1, Kingsville, VII.17.21; 2, Buffin Bay, Riviera Beach, VI.17.48 (S. S. Roback); Lee County, 1, Giddings, V.9.54 (L. D. Beamer); Reeves County, 1, Pecos, VI.1916 (R. H. and E. M. Painter); Runnels County, 2, Ballinger, VIII.2.31 (R. H. Painter); Uvalde County, 1, Sabinal, VI.13.10 (F. C. Scott).

Paravilla tricellula (Cole)

Fig. 48

Villa (Paravilla) tricellula Cole, 1952: 129; Painter and Painter, 1965: 439. (Type in California Academy of Sciences.). *Paravilla tricellula* (Cole), Hull, 1973: 392.

Known only from the type series plus four additional specimens. The species is easily recognized by the presence of three submarginal cells in the wings, the short, broad basal half of the third antennal segment, which is much wider than the second segment, and the yellowish thoracic and whitish abdominal hairs.

The original description was based on 17 supposedly female specimens. An examination of the type series disclosed a single male, which is described below.

This species is similar both to *P. pallida* and to *P. winburni*. It is distinguished from *P. pallida* by the three submarginal cells and from *P. winburni* by the presence of black hair on the side of the abdomen.

Male. - Black; first antennal segment, face, palpi, legs

except coxae and tips of tarsi fulvous.

Front with black hair and fulvous scales, scales reaching ocellar tubercle. First antennal segment short, not more than one and one-half times longer than second segment, expanded slightly on inner apical margin; second segment almost twice as wide as long, both basal segments with black hair; third segment one and one-half times longer than segments one and two combined, basal portion broad, conical, wider than second segment, apical styliform portion equal in length to basal portion. Face with short black hair and fulvous scales, a short line of fulvous hair below antennae. At most, tips of labellae projecting beyond oral margin. Palpus with black hair. Occiput whitish tomentose laterally, becoming yellowish toward vertex; cervical fringe yellow.

Mesonotum with short black hair on disk, pale-yellow hair on collar, whitish hair above root of wing to post-alar callus, tomentum hairlike, pale yellow, semi-erect, bristles yellow. Pleura pale-yellow to whitish pilose and tomentose. Coxae with yellow and black hair, scales whitish; fore and mid femora with black hair, yellow hair on hind femur, all femora with pale-yellow scales.

Halter stem fulvous, knob white. Squama brownish, with fringe of white scales. Scutellum with black hair, pale hair across base and at sides; tomentum whitish; bristles yellow.

Basicosta with yellow and black scales and black setulae. Wing hyaline, lightly infuscated with brownish yellow in costal cell; basal half of marginal, all of first basal, base of second basal, anal, and axillary cells colored lightly, discal cell hyaline except for small colored area at base, darker spots of color on r-m crossvein, at bases of third and fourth posterior cells; three submarginal cells present; vein at base of second posterior cell bent slightly.

Abdominal dorsum whitish pilose, becoming yellowish apically, long black hair across posterior halves of segments two to seven, small spot of black hair in middle of posterior margin of first segment, tomentum white to pale yellowish, darker apically; side of abdomen whitish pilose, tufts of black hair on apical corners of second and third segments, rest of margin mixed white and black-haired. Venter of abdomen with white hair at base, dark yellow hair from segment three to apex of abdomen, black hairs across posterior margins of segments four to seven, tomentum dark yellow to fulvous.

Genitalia with epandrium almost square, lower apical corner rounded, lower margin concave; epiphallus straight, apex enlarged, tip curved upward, apex and lower margin rounded; in dorsal view, epiphallus short, broad, apex rounded, sides concave near middle; aedeagus short, less than half length of epiphallus, broad; basistylus short, base and apex rounded; dististylus broad, small hook at apex.

Female. - Except for denser tomentum, does not differ appreciably from male.

Distribution. - Known only from southern California. The type series is from San Diego County and the four additional specimens are from Riverside and San Bernardino counties. All were collected in April and May.

The type series was reared from *Diadasia biturberculata* (Cresson).

Paravilla vigilans Coquillett

Fig. 49

Anthrax vigilans Coquillett, 1887: 176. (Type in U.S. National Museum.). *Villa (Paravilla) vigilans* (Coq.), Painter and Painter, 1965: 439. *Paravilla vigilans* (Cog.), Hull, 1973: 392.

Coquillett compared this species to *P. perplexa*, but it is more like *P. syrtis* than *P. perplexa*. It differs from *P. syrtis* by the absence of black hair and scales on the dorsum of the abdomen in the male. Some *P. syrtis* approach this condition, i.e., black or brown tomentum, when present on the abdominal dorsum, forms a median spot on segments two to four, giving the appearance of a vague mid-dorsal stripe, but in *P. syrtis* the black tomentum forms crossbands at least on the second segment. A point that emphasizes the closeness to *P. syrtis* is that not all females of *P. vigilans* can be separated from females of *P. syrtis*.

Male. - Black; basal antennal segments, most of face, legs except tarsi, posterior margins of ventral abdominal segments, last three tergites, genital capsule fulvous.

Front and face with black hair and pale-fulvous scales. Antennae with black hair; first segment two to three times longer than second segment, expanded on inner apical margin; second segment one and one-half times wider than long; third segment half again as long as segments one and two combined, basal half long conical, apical styliform portion equal to basal portion. At most, labellae projecting beyond oral margin. Palpus brown, with black hair. Side of occiput white to pale-yellow tomentose, darker-yellow hair and scales at vertex; cervical fringe pale yellow.

Mesonotum pale-yellow pilose, scattered black hair on disk, white hair above root of wing, tomentum thin, scattered, pale yellow to whitish, bristles pale. Pleura white pilose and tomentose. Fore and mid coxae with white and black hair, hind coxa white-haired, scales on all coxae whitish; fore and mid femora with black hair, absent on hind femur, scales on femora and tibiae pale yellow, bristles black.

Halter stem testaceous, knob white. Squama yellowish, with fringe of white scales. Scutellum with black hair, yellowish hair across base and at sides; tomentum yellowish, not dense; bristles pale.

Basicosta with pale-yellowish scales and black setulae. Wing hyaline, basal half or less infuscated with brown; margin of color runs diagonally basad from apex of subcosta across middle of discal cell below r-m crossvein to basal three-fourths of anal cell, margin of color including bases of third and fourth posterior cells, color in axillary cell not reaching hind margin of wing, both basal cells entirely colored, although second basal cell is lighter in color, faint spots of color at base of vein R_4 and at base of second posterior cell.

Abdominal dorsum entirely white pilose and tomentose. Venter white pilose and tomentose, with a few black hairs present on fourth and following segments.

Genitalia with epandrium more than twice as long as wide

at center, lower apical corner projecting slightly, rounded; lower margin straight; epiphallus curved slightly, broad, tip curved up, apex rounded, wrinkled, lower margin straight; in dorsal view, epiphallus ovoid in apical half, sides concave near middle; aedeagus about half length of epiphallus, curved downward; basistylus long, narrow, base rounded, apex more or less pointed; dististylus short, wide, hooked apically.

Female. - As described for male except for fulvous hair along eye margin on front, black hair across dorsal abdominal incisures, and denser tomentum.

Variation. - Some males have a small black tomentose spot on the second and sometimes the third dorsal abdominal segments; on occasion a few black hairs are present toward the apex of the abdomen. The spots of color at the bases of vein R_4 and the second posterior cell may vary from light to dark or even be absent, but the normal condition is as described. The more northernly specimens have black hair on the abdomen and the femora, or may on occasion be dusky.

Distribution. - Described from California. Additional specimens have been studied from Washington, Idaho, Wyoming, Utah, Colorado, Nevada, and British Columbia, Canada. The species is on the wing from June to September.

Paravilla winburni Hall, new species

Fig. 50

This species is similar to *P. tricellula*. The two species can be separated by the absence of black hair at the side of the second abdominal segment and the darker legs in *P. winburni*. The lower apical corner of the male epandrium is prolonged for a distance of about three times its medial width (lateral view).

Male. - Black; first antennal segment, oral margin, lower part of face, genital segments partially fulvous, legs dusky fulvous, almost ferruginous.

Front with black hair and yellowish to whitish tomentum, scales not reaching ocellar tubercle. First antennal segment twice length of second segment, expanded slightly on inner apical margin; second segment almost twice as wide as long, both basal segments with black hair, a few yellowish hairs also present on inner apex of first segment; third segment long, twice length of segments one and two combined, base long conical, apical styliform portion not quite equal in length to basal portion. Face with short black hair in middle above, yellow hair elsewhere, tomentum of pale-yellow scales. Proboscis not projecting. Palpus with black hair. Side of occiput white tomentose, patch of yellow scales behind ocellar tubercle; cervical fringe white.

Disk of mesonotum with short black hair, margins with pale-yellow hair, white hair above root of wing, tomentum dense, pale yellow, patch of whitish tomentum between humeral callus and wing base and across posterior margin of mesonotum, bristles yellow. Pleura white to very pale-yellow pilose and tomentose. Coxae with white hair and scales, hair at apex of fore coxa yellowish; fore and mid femora with black hair, absent on hind femur, tomentum on femora pale yellow to blackish, translucent; tibiae with pale scales, setulae and bristles black.

Halter stem yellow, knob yellowish white. Squama whitish, with fringe of white scales. Scutellum with short black hair medially, yellow to white hair at sides; tomentum pale yellow, white at sides, at base and along posterior margin; bristles yellow.

Basicosta with yellow scales and black setulae. Wing hyaline, with light-brown color in costal, subcostal, basal half of marginal, first basal, basal half of discal cells to r-m crossvein, lighter color in second basal and basal half of anal cells, axillary cell hyaline; three submarginal cells present; vein at base of second posterior cell sinuous.

Abdominal dorsum white to pale-yellow pilose, hair not dense, black hair across posterior margins of segments two to five; tomentum pale yellow, with white tomentum across basal half of second segment and on all of seventh segment, middle of segments two to four with a spot of dark brown to black scales; lateral margin of abdomen white pilose at base, becoming pale yellow apically. Venter of abdomen white pilose and tomentose.

Genitalia with epandrium narrow, lower apical corner prolonged for a distance about three times median width of prolongation, lower margin straight but deeply incised; epiphallus arched, tip curved upward, apex rounded, lower margin convex; in dorsal view, epiphallus spatulate, apex rounded, sides narrowed basally; aedeagus short, half length of epiphallus, pointed downward; basistylus broad, apex narrowed;

dististylus long, apex not strongly hooked.

Female. - Very similar to male, but with first antennal segment and legs a little darker and black hair on abdominal dorsum more prominent.

Variation. - In some specimens the scales on the femora are quite dark; in others they are light; in either case, they are somewhat translucent. Not all specimens have the small brownish spots of tomentum in the middle of abdominal segments two to four. In most of the specimens the wing coloring is very light; in a few it is dark brown. In the latter case, the wing is more extensively colored.

Holotype male and allotype. - From Tucson, Pima County, Arizona, VIII.9.30 (T. F. Winburn, R. H. Painter). Both in collection of U.S. National Museum.

Paratypes. - UNITED STATES: Arizona: Cochise County, 1, 12 mi. S Dos Cabezas, IX.5.58 (F. G. Werner); 3, Wilcox, VIII.11.30 (T. F. Winburn, R. H. Painter); Gila County, 1, Globe, VII.27.32 (R. H. Painter); Pima County, 6, topotypic, collected with type; 6, 36 mi. SW Tucson, VIII.3.32 (R .H. Painter); Yavapai County, 1, Congress Junction, IX.17.64 (F. M. and M. C. Hull). New Mexico: Hidalgo County, 1, Steins,, VIII.8.32 (R. H. Painter); 3, 18 mi. N Rodeo, VIII. 25.58 (W. R. Brown, R. M. Bohart); Otero County, 1, 8 mi. W Mescalero, IX.11.62 (R. H. and E. M. Painter). Texas: Reeves County, 2, Pecos, VI.16 (R. H. and E. M. Painter); Findlay, VIII.11.30 (T. F. Winburn, R. N. Painter). MEXICO: Sonora: 1, 59 mi. S. Hermosillo, VIII.13.59 (R. H. and E. M. Painter).

Paravilla xanthina (Painter)

Fig. 51

Villa (Paravilla) xanthina Painter, 1933: 13; Painter and Painter, 1965: 439. (Type in U.S. National Museum.).
Paravilla xanthina (Painter), Hull, 1973: 392.

This species, together with *P. nigronasica*, is one of the more easily recognized species of *Paravilla*. The wings are almost entirely hyaline and the body is yellow to bright-orange pilose, with crossbands of black tomentum on at least abdominal segments two to four. In the male, the tip of the abdomen is white tomentose.

Male. - Large size, length 10-14 mm. Black; first two antennal segments, face, legs except basal half or more of femora, especially fore femur and tips of tarsi fulvous.

Front black or fulvous pilose, usually mixed; tomentum yellow. Antennae with black hair, first segment twice length of second segment, cylindrical, not appreciably expanded on inner apical margin; second segment slightly wider than long; third segment about one and one-half times longer than segments one and two combined, basal portion long conical, apical styliform portion shorter than basal portion. Face yellow pilose and tomentose, sometimes with a few scattered black hairs also present. Proboscis not projecting beyond oral margin. Palpus with black and yellow hair. Side of occiput with pale-yellow scales, darker scales and short hair on vertex; cervical fringe pale yellow.

Mesonotum fulvous pilose and tomentose, hair slightly lighter in color above root of wing, bristles yellow. Pleura fulvous pilose and tomentose. Coxae fulvous pilose and tomentose, with a few black hairs on fore coxae near apex; fore and mid femora with black hair, absent on hind femur, scales on femora and tibiae yellow, bristles black.

Halter stem fulvous, knob white. Squama fulvous with fringe of pale-yellow scales. Scutellum fulvous pilose and tomentose; bristles fulvous.

Basicosta with fulvous scales and black setulae. Wing hyaline; faint color present in costal, first basal, basal half of second basal cells and at extreme base of wing, slightly darker spots of color on r-m crossvein and at base of fourth posterior cell; vein at base of second posterior cell more or less sinuous.

Abdominal dorsum fulvous pilose and tomentose, black tomentose crossbands on posterior halves of segments two to four, bands not reaching lateral margin of segments, sixth and seventh segments with white hair and scales; lateral margin of abdomen entirely fulvous pilose except for white hair on two terminal segments. Venter of abdomen fulvous pilose and tomentose.

Genitalia with epandrium a little longer than median width, lower apical corner rounded, lower margin straight or nearly so; epiphallus straight, thick, tip curved up, apex rounded, lower margin convex; in dorsal view, epiphallus rounded apically, sides subparallel, concave below middle; aedeagus short, less than half length of epiphallus, curved downward; basistylus broad basally, apex narrowed, rounded; dististylus long, narrow, small hook at apex.

<u>Female</u>. - Front entirely fulvous pilose; palpi with black

hair; terminal abdominal segments fulvous pilose and tomentose. Otherwise as described for male.

Variation. - The color of the body pile varies from light yellow to dark fulvous. Occasionally a male will have the frontal hair all fulvous as in the female; some females have black hairs on the front. The fore femora are normally black, with only small areas on the mid and hind femora dusky, but there is a wide variation in the amount of dusky or black color present.

Distribution. - This is a midwestern species, occurring from South Dakota to Nebraska and Texas in June and July. The Catalog of Diptera of North America lists specimens from Arizona but I have not seen any from that state.

Literature Cited

Bezzi, M.

1924. *The Bombyliidae of the Ethiopian Region.* British Museum (Natural History), London. 390 pp.

Bigot, J. M. F.

1892. "Diptéres nouveaux ou peu connus. 37th partie. XLVI: Bombylidi (mihi)." Soc. Ent. de France Ann. 61:321-376.

Bowden, J.

1964. "The Bombyliidae of Ghana." Ent. Soc. Southern Africa Mem. no. 8:1-159.

1975. "Bombyliidae." Vol. II in *A Catalogue of the Diptera of the Oriental Region.* University of Hawaii Press, Honolulu. 459 pp.

Cole, F. R.

1923. "Expedition of the California Academy of Sciences to the Gulf of California in 1921. The Bombyliidae (bee flies); Diptera from the islands and adjacent shores of the Gulf of California." II. Calif. Acad. Sci. Proc. Ser. 4, 12 (13):289-314.

1952. New Bombyliid flies reared from Anthophorid bees

(Diptera: Brachycera). Pan-Pacific Ent. 28:126-130.

Coquillett, D. W.

1887. "Monograph of the genus *Anthrax* north of Mexico." Amer. Ent. Soc. Trans. 19:168-187.

1894. "Notes and descriptions of North American Bombyliidae." Amer. Ent. Soc. Trans. 21:89-112.

1902. "New Orthorraphous Diptera from Mexico and Texas." N. Y. Ent. Soc. Jour. 10:136-141.

Engel, E. O.

1932-1937. "Bombyliidae." Vol. 4, pt. 3. in E. Lindner, ed., *Die Fliegen der palaearktischen Region.* Schweizerbart publ., Stuttgart. 619 pp.

Graenicher, S.

1910. "Some new and rare Diptera from Wisconsin." Canad. Ent. 42:26-29.

Hall, J. C.

1970. "The North American species of *Villa (Thyridanthrax)* (Osten Sacken). (Diptera: Bombyliidae)." Amer. Ent. Soc. Trans. 96:519-542.

1976 (1975). "The Bombyliidae of Chile (Diptera: Bombyliidae)." Univ. Calif. Pub. Ent. 76:1-278.

Hesse, A. J.

1956. "A revision of the Bombyliidae (Diptera) of Southern Africa." Part III. Southern Africa Mus. Ann. 35:465-972.

Hull, F. M.

1973. *Bee flies of the world. The genera of the Family*

Bombyliidae. Smith. Inst. Press, Washington, D.C. 687 pp.

Jaennicke, F.

1867. "Neue Exotische Dipteren." Senckenb. Naturf. Gesell. Abhandl. 6:311-408.

Loew, H.

1869. "Diptera Americae Septentrionalis Indigena. Centuria octava." Berlin Ent. Zeit. 13:1-52.

Macquart, J.

1846. "Diptéres exotiques nouveaux ou peu connus." 1st Supplement. Soc. Roy. Sci. Agr. et des Arts, Lille, Mem. 1844:133-364.

Osten Sacken, C. R.

1886. "Diptera." Vol. I. [in part] in F. D. Godman and O. Salvin, eds., *Biologia Centrali-Americana*. Zoologia-Insecta-Diptera, London. pp. 75-128.

Painter, R. H.

1933. "New subgenera and species of Bombyliidae (Diptera)." Kansas Ent. Soc. Jour. 6:5-18.

Painter, R. H., and E. M. Painter.

1962. "Notes and redescriptions of types of North American Bombyliidae (Diptera) in European museums." Kansas Ent. Soc. Jour. 35:2-164.

1965. "Bombyliidae." In Stone et al., eds., *A Catalog of the Diptera of America north of Mexico*. Agric. Res. Serv., U. S. Dept. Agric., Washington, D.C. pp. 407-446.

1974. "Notes on, and redescriptions of, types of South

American Bombyliidae (Diptera) in European and United States museums." Kansas State Agric. Exp. Sta. Res. Pub. 168, 322 pp.

Painter, R. H., E. M. Painter, and J. C. Hall.

1978. "A catalogue of the Diptera of the Americas south of the United States." Mus. Zool. Univ. São Paulo, Brazil, fasc. 38. pp. 1-92.

Walker, F.

1849. "List of the specimens of dipterous insects in the collection of the British Museum." BMNH, London, 4:689-1172.

1852. "Diptera." In W. W. Saunders, ed., *Insecta Saundersiana*, London, pp. 157-414.

Wiedemann, C. R. W.

1830. *Aussereuropäische zweiflügelige Insekten*, vol. 2. Hamm, Vienna. 684 pp.

Williston, S. W.

1901. Supplement [part]. In F. D. Godman and O. Slavin, eds., *Biologia Centrali-Americana*. Zoologia-Insecta-Diptera, London. pp. 1-378.

Index to Figures

Figures 1 through 51: male genitalia of *Paravilla* species. (a) lateral view of genital structures; (b) dorsal view of epiphallus; (c) posterior view of apex of epiphallus; (d) lateral view of epandrium. All figures were drawn with the aid of a camera lucida and all were made at the same magnification.

Fig. 1 *acutula*
Fig. 2 *albata*
Fig. 3 *albicera*
Fig. 4 *aridula*
Fig. 5 *borea*
Fig. 6 *californica*
Fig. 7 *castanea*
Fig. 8 *cinerea*
Fig. 9 *consul*
Fig. 10 *cunicula*
Fig. 11 *deserta*
Fig. 12 *diagonalis*
Fig. 13 *edititoides*
Fig. 14 *eminens*
Fig. 15 *emulata*
Fig. 16 *epheba*
Fig. 17 *extremitis*
Fig. 18 *flavipilosa*
Fig. 19 *floridensis*
Fig. 20 *fulvicoma*
Fig. 21 *fumida*
Fig. 22 *fumosa*
Fig. 23 *hulli*
Fig. 24 *imitans*
Fig. 25 *inatra*
Fig. 26 *lacunaris*
Fig. 27 *leucothoa*
Fig. 28 *lucida*

Fig. 29 *macneilli*

Fig. 30 *mercedis*

Fig. 31 *mexicana*

Fig. 32 *montivaga*

Fig. 33 *nigriventris*

Fig. 34 *nigrofemorata*

Fig. 35 *nigronasica*

Fig. 36 *opaca*

Fig. 37 *painterorum*

Fig. 38 *palliata*

Fig. 39 *pallida*

Fig. 40 *parasitica*

Fig. 41 *perplexa*

Fig. 42 *parvula*

Fig. 43 *separata*

Fig. 44 *spaldingi*

Fig. 45 *splendida*

Fig. 46 *syrtis*

Fig. 47 *texana*

Fig. 48 *tricellula*

Fig. 49 *vigilans*

Fig. 50 *winburni*

Fig. 51 *xanthina*

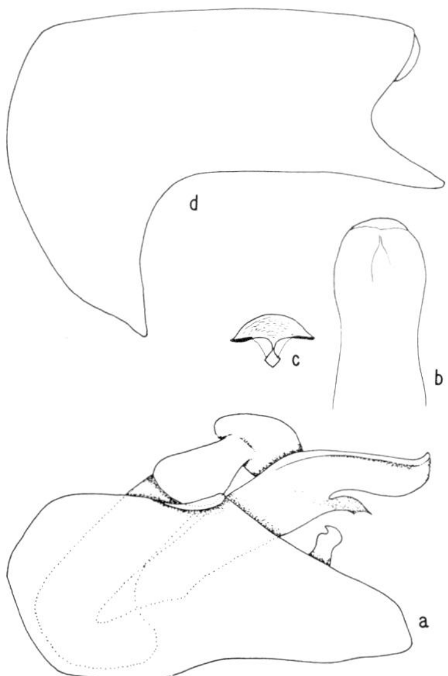

1 ACUTULA

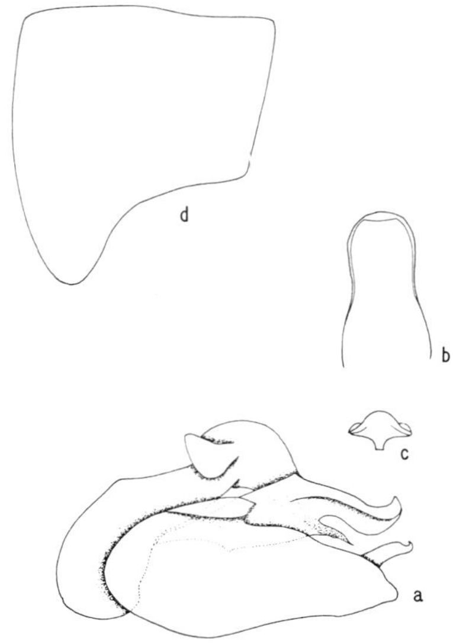

2 ALBATA

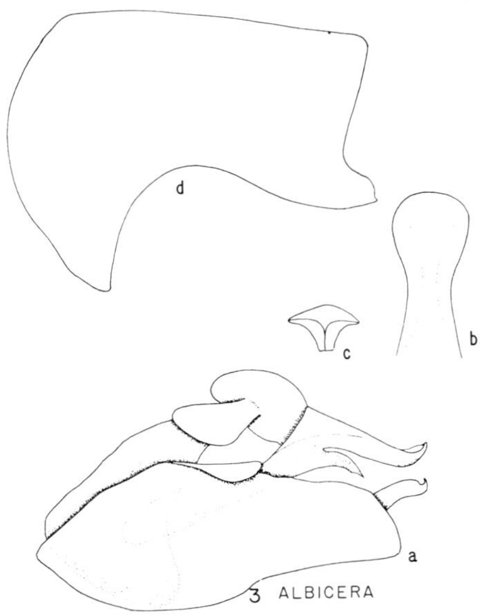

3 ALBICERA

4 ARIDULA

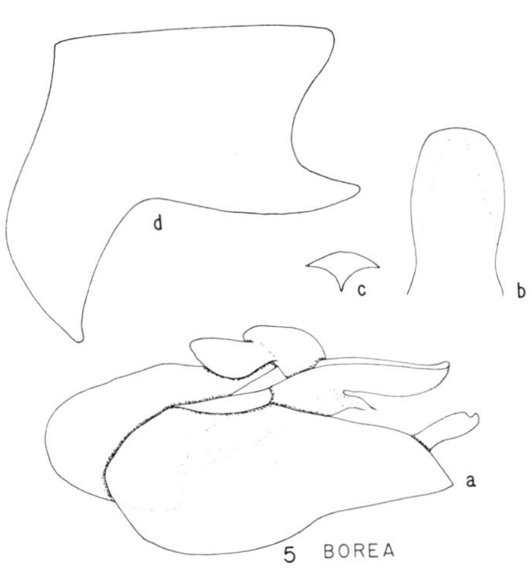

5 BOREA

6 CALIFORNICA

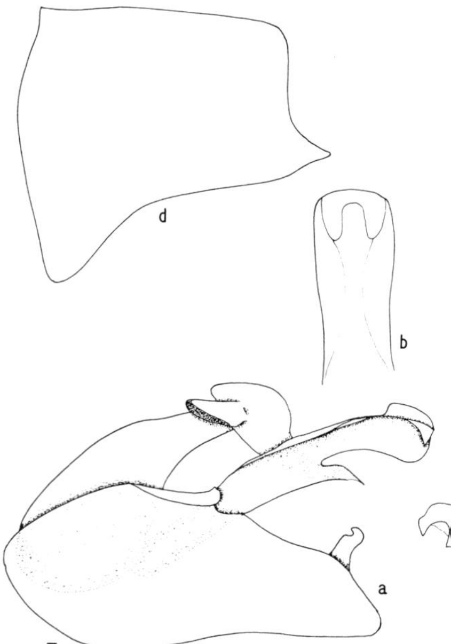

7 CASTANEA

8 CINEREA

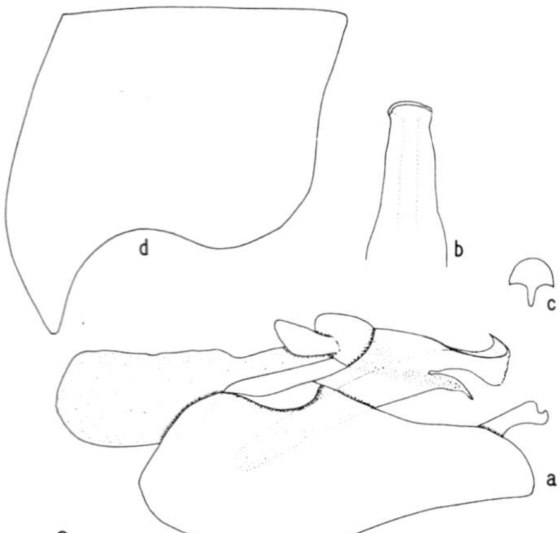

9 CONSUL

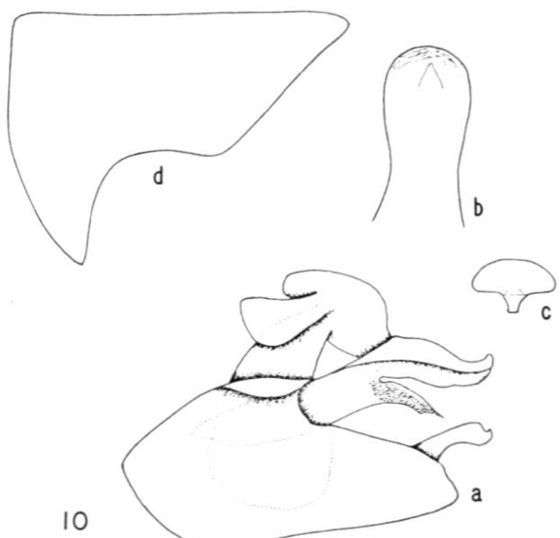

10 CUNICULA

11 DESERTA

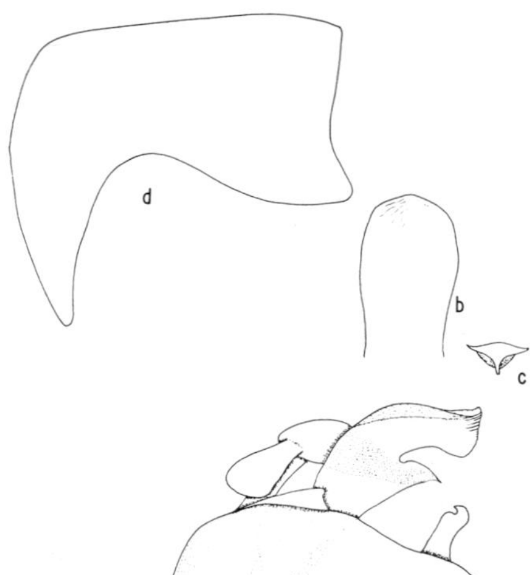

12 DIAGONALIS

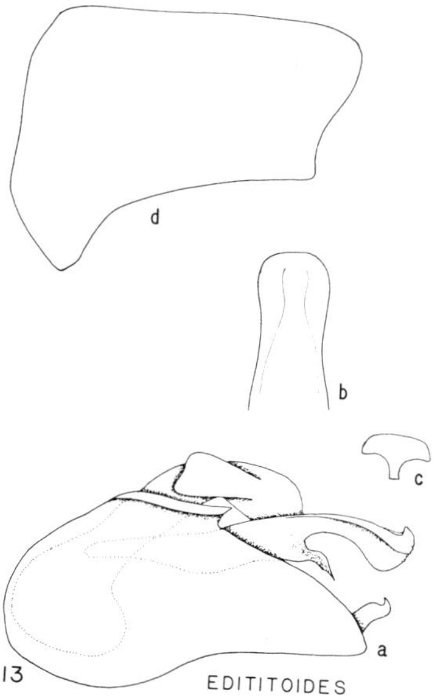

13 EDITITOIDES

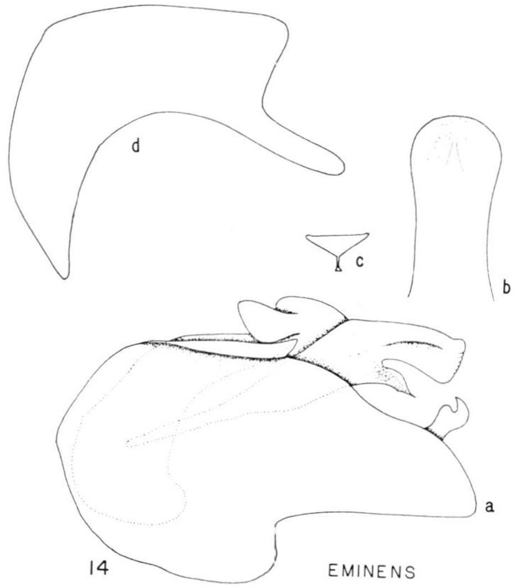

14 EMINENS

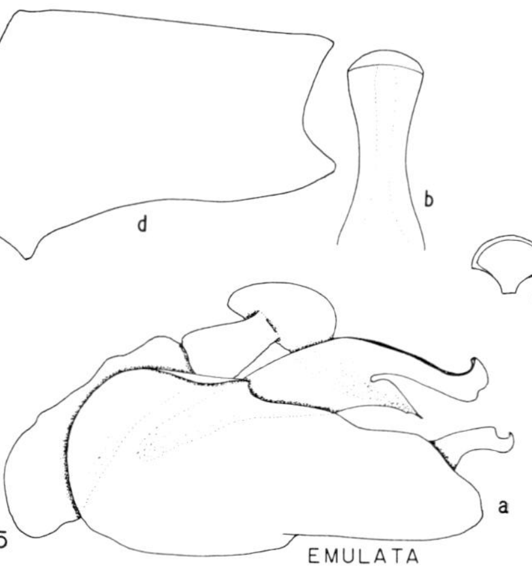

15 EMULATA

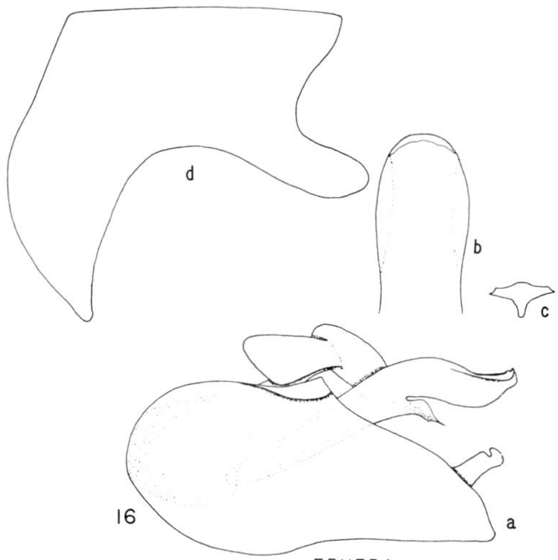

16 EPHEBA

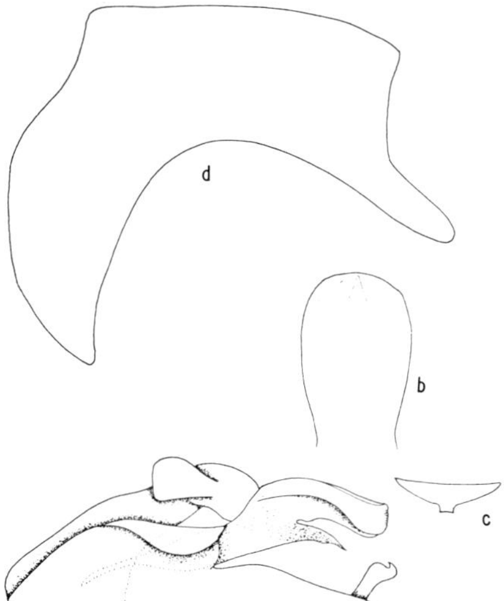

17 EXTREMITIS

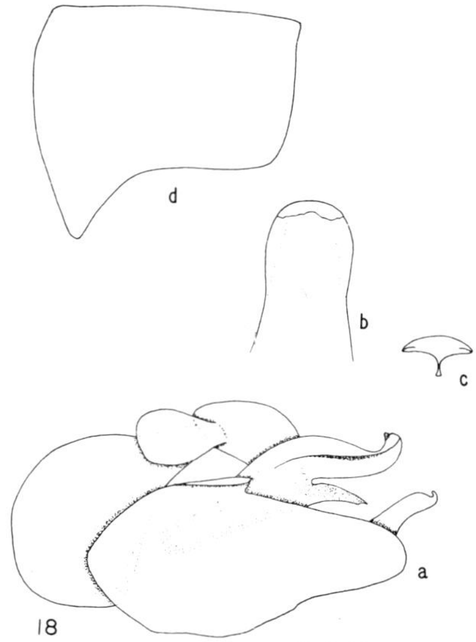

18 FLAVIPILOSA

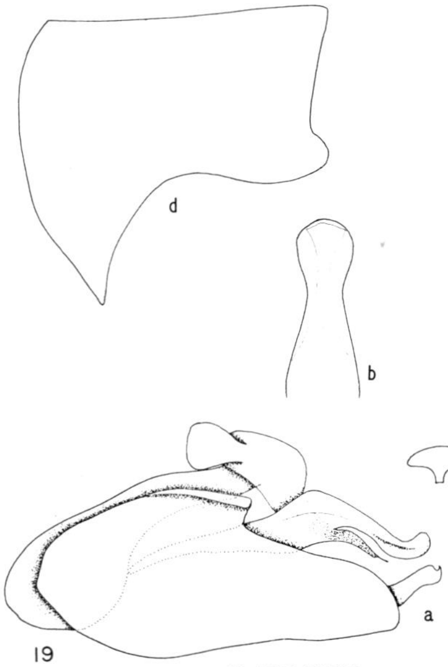

19 FLORIDENSIS

20 FULVICOMA

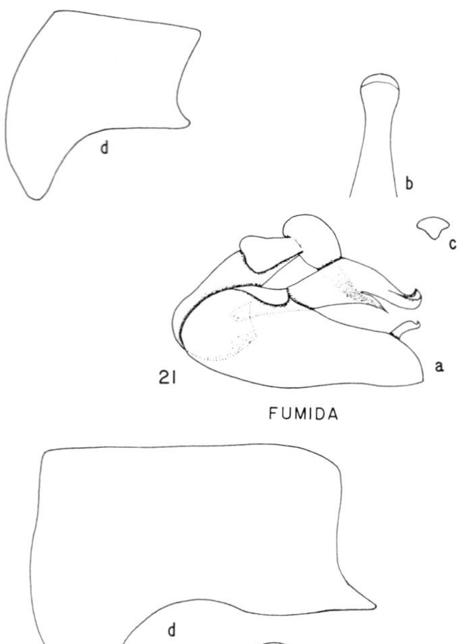

21 FUMIDA

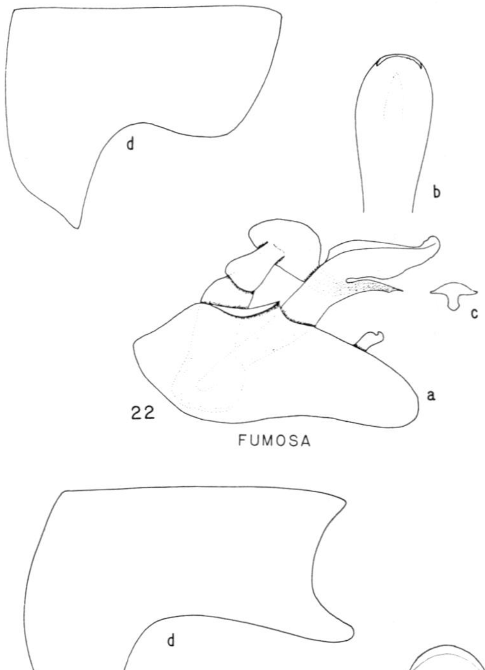

22 FUMOSA

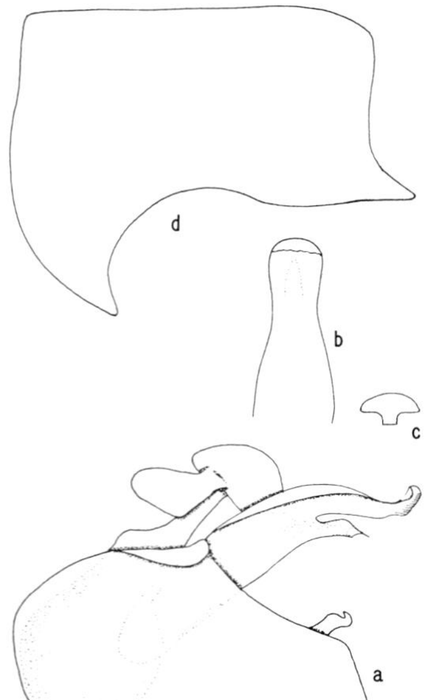

23 HULLI

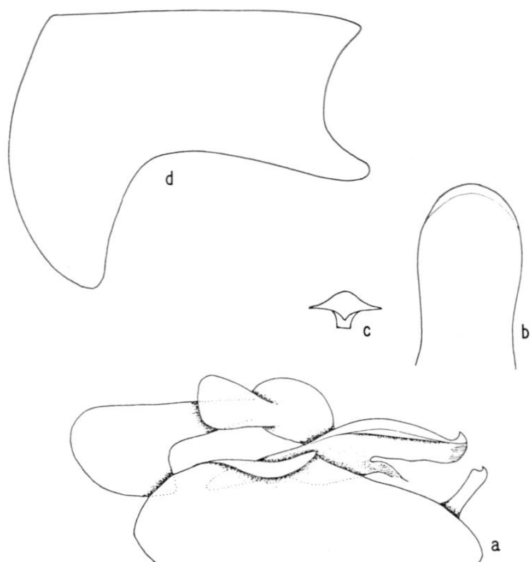

24 IMITANS

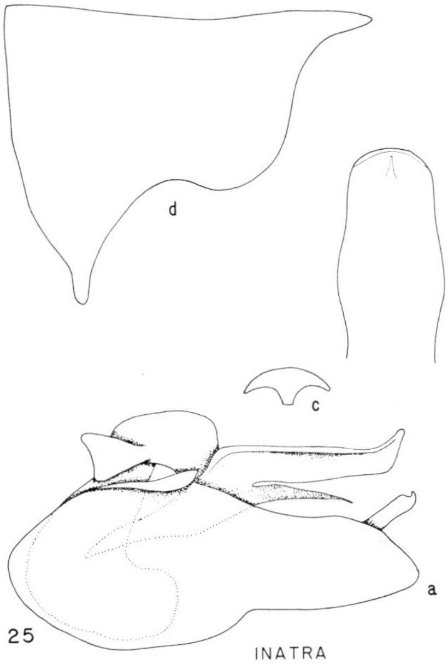

25 INATRA

26 LACUNARIS

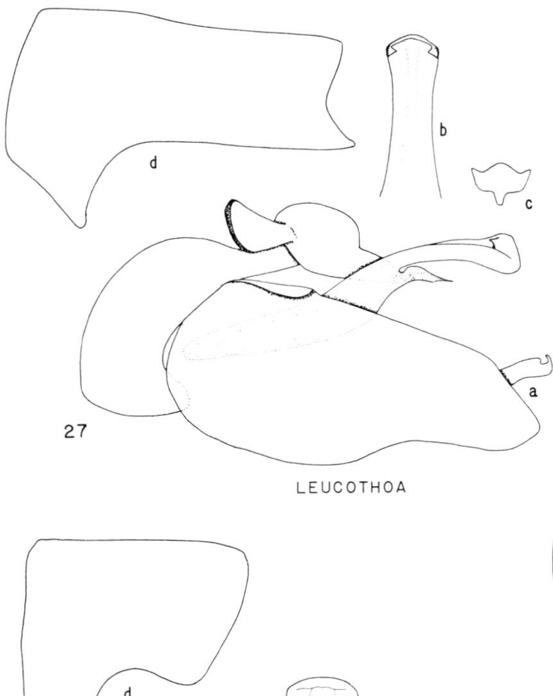

27 LEUCOTHOA

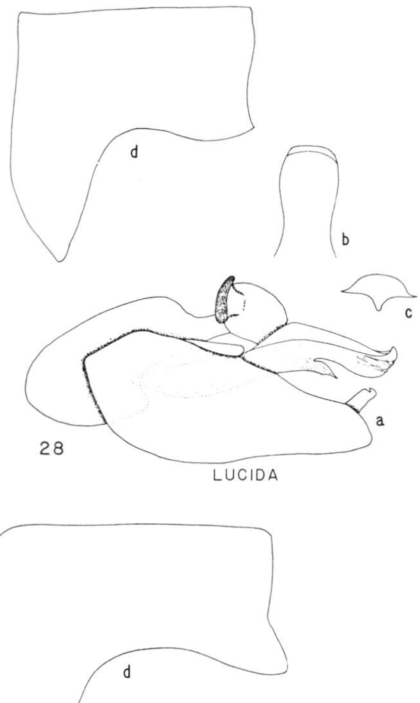

28 LUCIDA

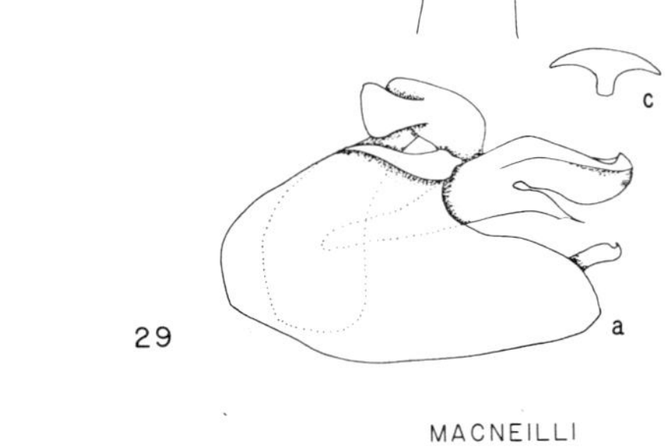

29 MACNEILLI

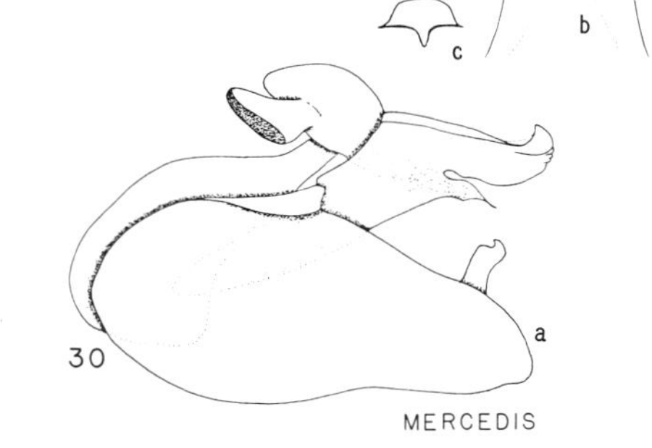

30 MERCEDIS

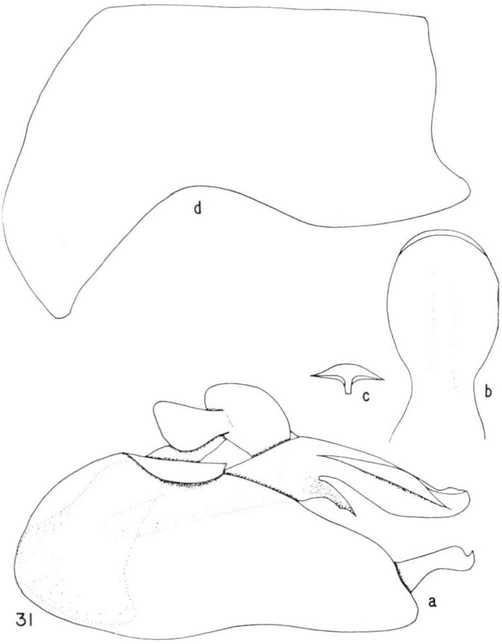

31 MEXICANA

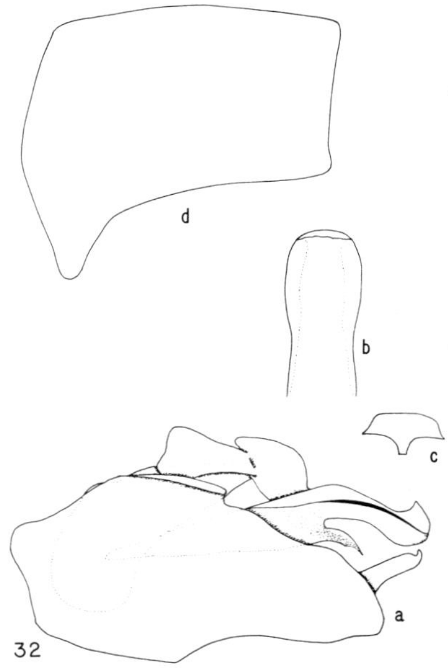

32 MONTIVAGA

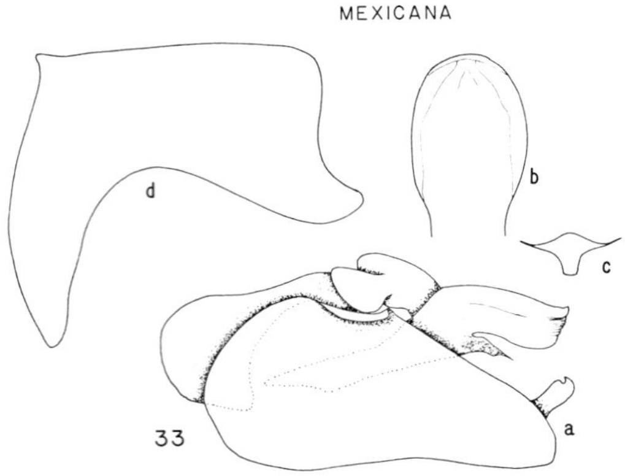

33 NIGRIVENTRIS

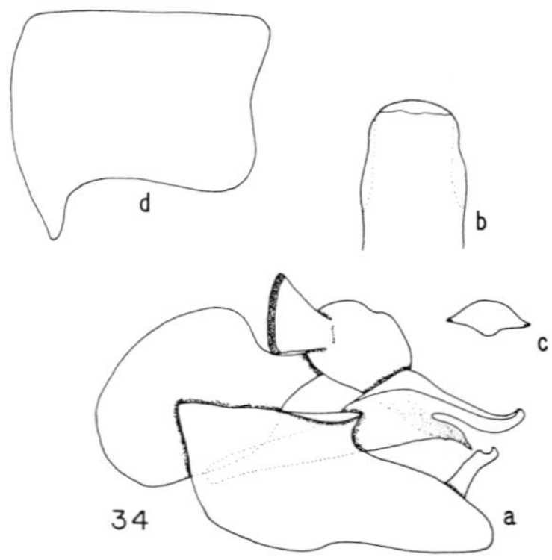

34 NIGROFEMORATA

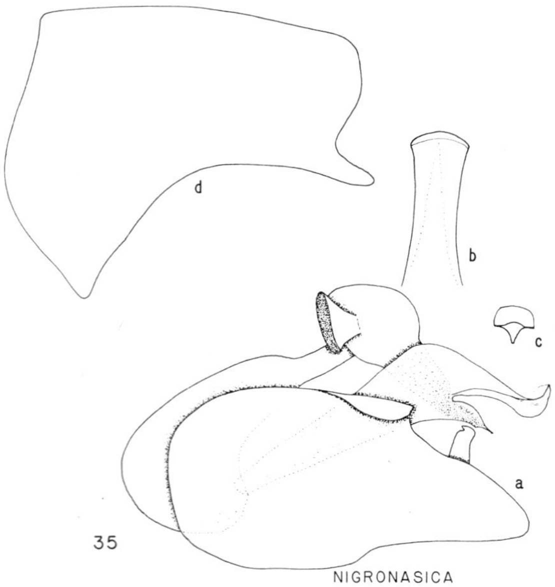

35 NIGRONASICA

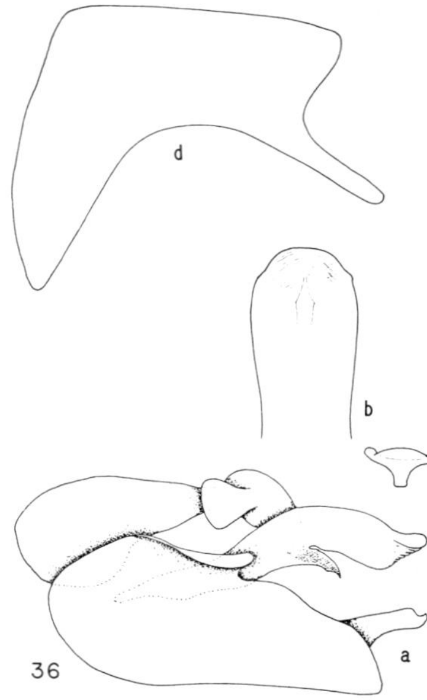

36 OPACA

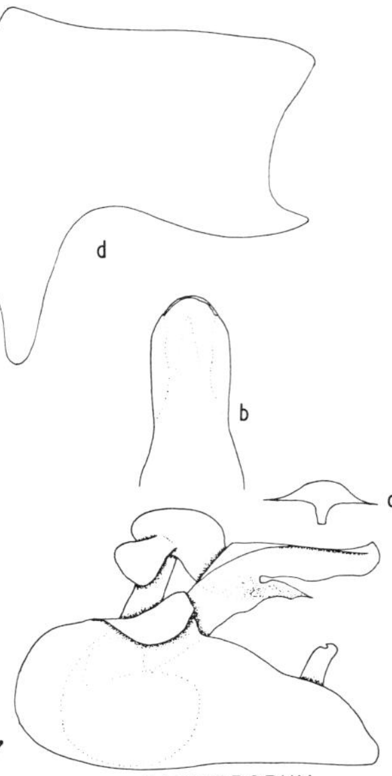

37 PAINTERORUM

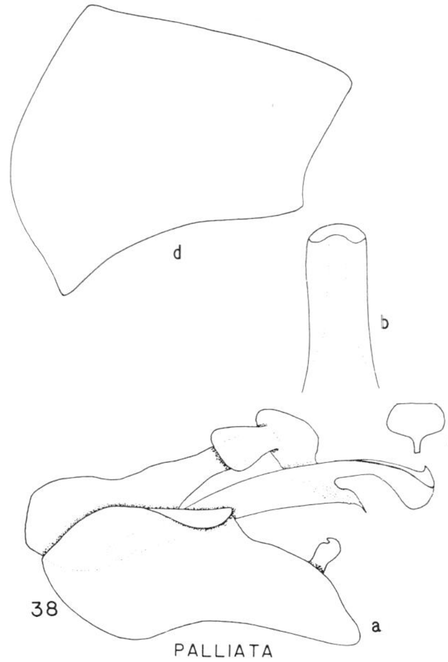

38 PALLIATA

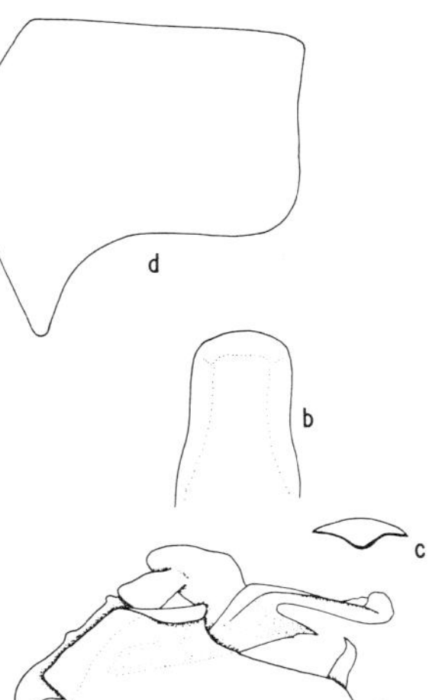

39 PALLIDA

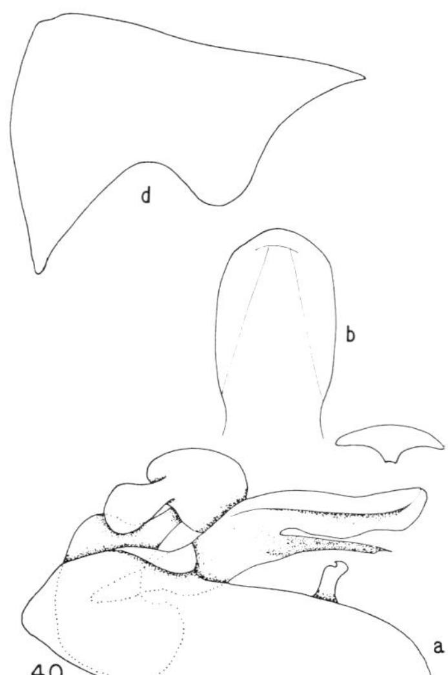

40 PARASITICA

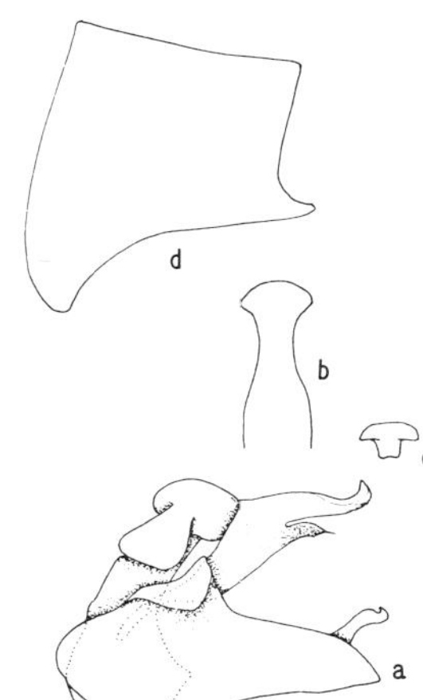

41 PERPLEXA

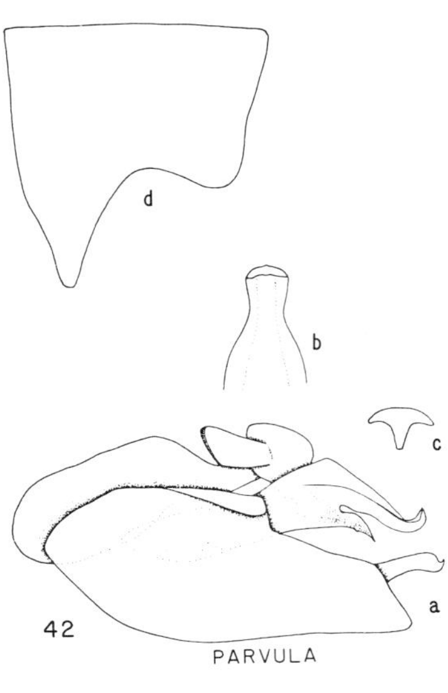

42 PARVULA

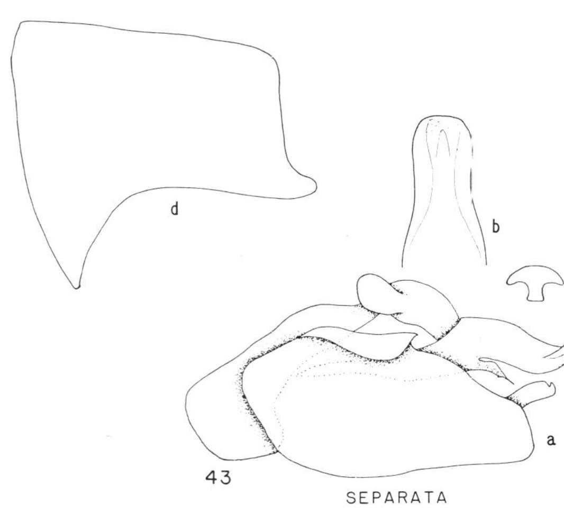

43 SEPARATA

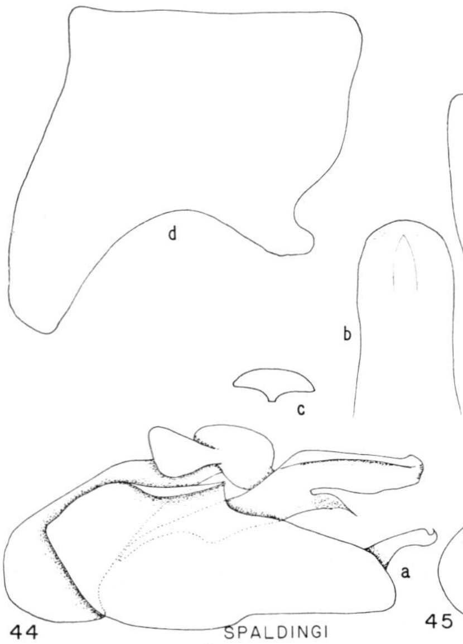

44 SPALDINGI

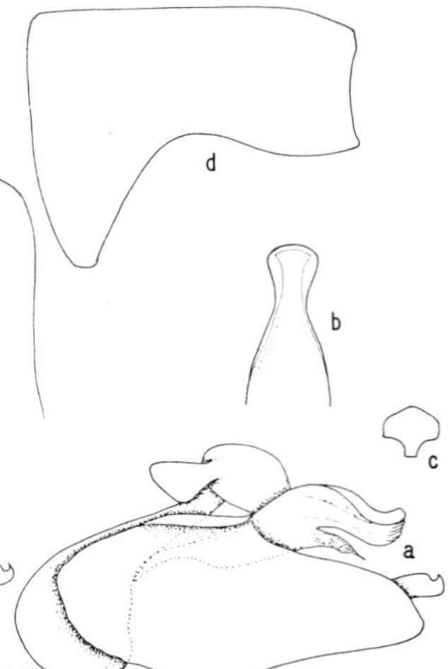

45 SPLENDIDA

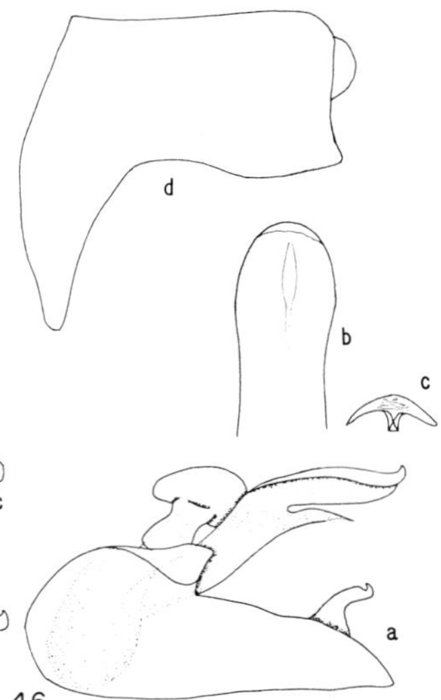

46 SYRTIS

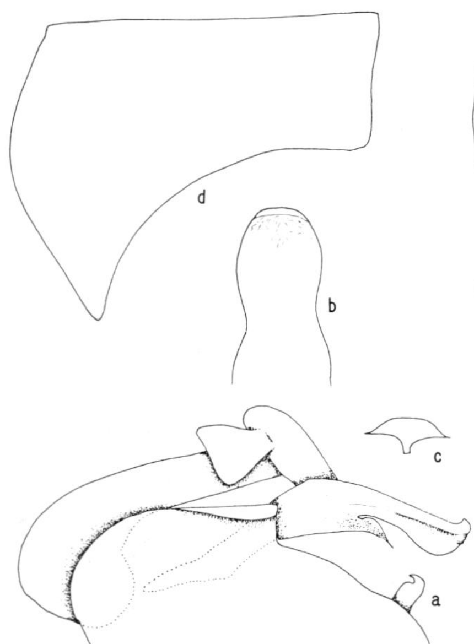

47 TEXANA

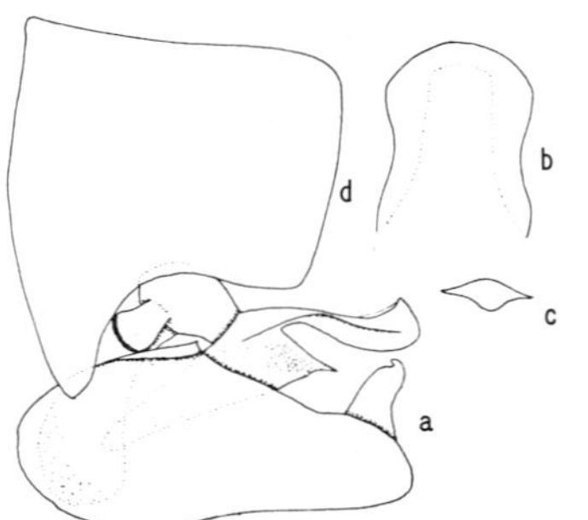

48 TRICELLULA

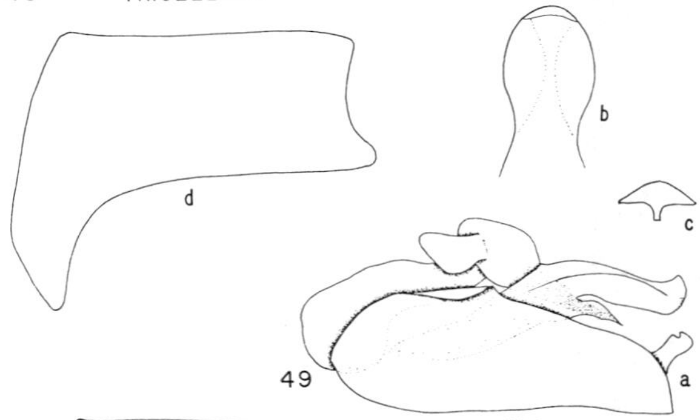

49 VIGILANS

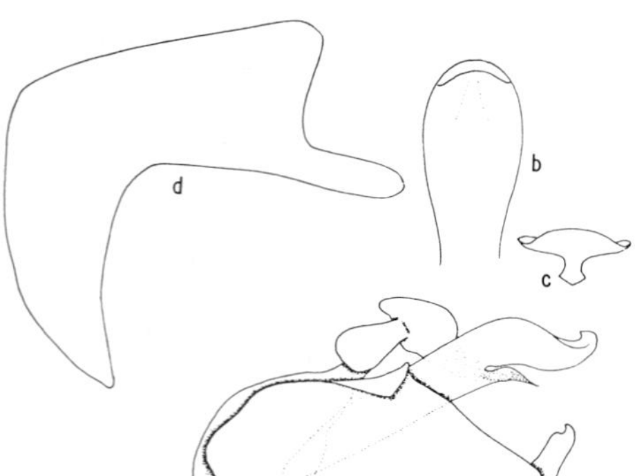

50 WINBURNI

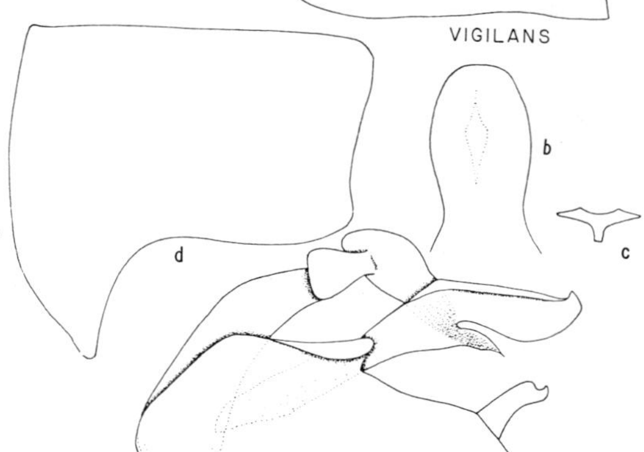

51 XANTHINA